# An Illustrated Glossary of Botanical Terminologies
## *(An Easy Approach to Plants Terms)*

## *(First Edition)*

### Authored By

### Hasnain Nangyal

*Department of Botany*
*Faculty of Sciences*
*Hazara University*
*Mansehra*
*Khyber Pakhtoonkwa*
*Pakistan*

# DEDICATION

*My Beloved Grand Father Syed Badshah*

*&*

*Father Syed Bahadur*

# CONTENTS

# About the Author

**Mr. Hasnain Nangyal** is a research fellow Department Of Botany, Faculty of Sciences Hazara University, Mansehra, Khyber Pakhtoonkhwa. He has published remarkable review articles and research papers in well reputed national, international scientific impact factor journals, magazines and newspapers like Technology Times, Microbiology World, *etc*. He is an active member of many national and International Research Organisations, *e.g*, Exective member of Pakistan Medical Microbiology Association, Member Pakistan Medicinal Plants Association, Member North Carolina Herbs Associatin (NCHA) USA. Ambassador Pakistan Society of Psycophysiology, Member of Editorial Board Member of Scientific Magazine "Microbiology World" a bimonthly magazine ISSN no 2350-8744 Published from Nepal, Member of Editorial Board Advances in Biomedicine and Pharmacy (An International Journal of molecular medicine and pharmacy) Published from UK, Member of Editorial Board International Journal of Endorsing Journal of Endorsing Health Science Research Impact Factor 3.6 Indexed by Eastern Mediterian Index Medicus –WHO, Member of Editorial Board Journal of Biosciences and Agricultural Research ISSN no 2312-7945 Published from Bioscience Information Network Bangladesh, Reviewer FUUAST Journal of Biology, Joint Sectary Pakistan Herbs Society a sister organization of NCHA (North Carolina Herbs Association) North Carolina USA, Member Working Group "By Science it will be a paradise (Scientific Group)" Cairo Egypt, Life time Membership Biocognizance, and a Member of Editorial Board "Paradise of Science" a bimonthly multidisciplinary Magazine published from Cairo University, Cairo, Egypt.

He attended International Scientific Conference of young Scientists entitled "The role of multidisciplnary approach in solution of actual problems of fundamental and applied Sciences (Earth, Chemical, Technical)" in October 2014 in National Academy of Sciences Azerbaijan, and was invited as a key note speaker by the Academy of Sciences Azerbaijan

He attended Indian National Science Congress held in Bombay University on January 10, 2015 and was invited as a key note speaker by Inidan National Science.

He attended International Symposium "Biotechnology Recent Advances" Organized by Department of Biotechnology University of Malakand Khyber Pakhtoonkhwa in 2008.

## From Head of Department

I am privileged to record that Mr. Hasnain Nangyal has taken initiative to write a book on Botanical terminologies which is the need of the hour. I foresee a very bright future for the author who is young with new and creative ideas. Best of luck.

*Manzoor Ahmed*
Department of Botany
Hazara University Mansehra
Khyber Pakhtoonkhwa

# FOREWORD

I am always reluctant to single out of some particular feature of the work of a researcher because it tends the inevitably to simplify and reduce the work. But in this book about Glossary of Botanical Terminologies by Hasnain Nangyal, it should not be out of place to observe that he has very rare ability to highlight his ideas rather than just talking about them. By making their points through his own unique way he gain the added power of allowing readers what is going on rather than what is told. He do this with such a dazzling skill, you never see the ideas coming and don't realize until much later how profoundly they have reached your mind.

*Vladimir V. Titok*
The Central Botanical Garden State Scientific Institution Suranov Minsk
The National Academy of Sciences of Belarus
Republic of Belarus

# PREFACE

Terminology of Plants is one of the most fascinating scientific disciplines and it has immense importance. It has been advancing repeatedly and got different potential applications due to the surge of usefulness of plants. The present glossary is intended for beginners who wish to understand the fundamentals of plant Science terminology. The whole gamut of subject has been described in simple and systematic manner and suites to the readers who have not been exposed sufficiently to the subject. The primary aim throughout has been clarity, simplicity and high standard.

This glossary is complied to provide adequate information of morphological, anatomical, taxanomic, and other vital aspects of plants suggested by undergraduate students. Every effort has been made to make the book useful for the students, teachers and research scholars in this field.

The author has freely consulted various articles, discussion notes, reviews, extracts and published glossaries from various countries in the preparation of this book, in order to make it comprehensive and up-to-date,and terms are explained in such a way that a reader can pick it easily. Unnecessary details have been avoided.Though every care has been taken by me, publisher and the printer, it is quite likely that some errors might occur in the book, but I hope these are of very insignificant nature. Moreover, explanatory positive criticism and concrete suggestions will be appreciated gratefully.

The author expresses gratitude to his friends and colleagues whose constant inspirations have initiated him in bringing out his book.

Special thanks to my Supervisor and staff of Bentham Publishers for their whole hearted cooperation in the publication of this book

*Hasnain Nangyal*
Department of Botany
Faculty of Sciences
Hazara University
Mansehra
Khyber Pakhtoonkwa
Pakistan

# ACKNOWLEDGEMENTS

I am very grateful to my advisory panel for their valuable comments, suggestions additions and worthwhile improvements without them I would not have been able to do this work so easily.

My research Supervisor Dr Azhar Hussain Shah, Faculty member Department of Botany, Faculty of Life Sciences Hazara University, Mansehra, Khyber Pakhtoonkhwa.

Mr. Ram Bahadur Thapa Director General Department of Plant Resources Ministry of Forests and Soil Conservation Government of Nepal.

Dr. R.K Chakravety, Director Acharya Jagadish Chandra Bose Indian Botanic Garden, Shibpur Howrah, West Bengal.

Kate Sackman, Member of Lake Forest Open Lands Association and Technology Member Botanical Gardens Conservation International.

James M. Affolter, Proffesor and Director of research the state Botanical Garden of Georgia, Milledge Ave, Athens, Greece.

Santosh Kumar, Conservator of Forests and Chief Wild Life Warden Union Terrioritry, Chandigarh, India.

Dr. P.G Latha, Director Jawaharlal Nehru Tropical Botanic Garden and Research Institute Pacha Palode Thiruvananthapuram, Kerala, India.

Dr. Gohar Oganesova, Vice President Armenian Botanical Society, Armenia.

Jeff Chandler Senior Technician (Horticulture) Andromeda Botanical Gardens University of West Indies.

Dr. Vladimir V. Titok, Director The Central Botanical Garden State Scientific Institution Suranov Minsk, The National Academy of Sciences of Belarus, Republic of Belarus.

Dr. Director: Aleksandr V. Pugachevskii, The V.F Kuprevich Institute of Experimental Botany Akademichnaya Street, The National Academy of Sciences of Belarus Minsk, Republic of Belarus.

Dr. Keith Bensusan, Director the Alameda Gibraltar Botanic Garden Gibraltar, Spain.

Dr. Sefidkon, Fateme Deputy of Research Services Research Institutes for Forest & Rangelands, Mashad, Iran.

# INTRODUCTION

Every author efforts are taken admiringly or critically with the qualities of his or her subject matter and shortcomings, also as we all know, everybody tries his level best to enlighten reader's mind with some innovative ideas of thinking his subject matter in a different context, as i have not experienced any discouragement in any circle in this regard, and hence I manage to write a glossary of botanical terms.

The purpose to write a glossary on this very subject was that, I found myself using the available glossaries available in the market. However, I found them a little bit outdated and hard to study specially for Undergraduate students, because they did not have proper background at their stage, so then I thought that a handy a easily understandable glossary is the need of the day specially for Undergraduate students and general science students who study Botany as one of their subject (Pharmacy, Microbiology, Biochemistry and Pharmacognocy *etc.*). This current glossary has still based on standard glossaries available in the market but it have the remarkable characters that it is short, comprehensive and not too much complex moreover I enlarged some terminologies by existing botanical works simply by re analyzing it, it have 1552 terms and 244 Illustrations, (Pictures are classified into two categories here Coloured and simple coloured pictures are provided by James G.Harris and Melinda Woolf Harris writers of Plant Identification Technology and Simple pictures were taken from "Principles of Botany" written by Tanweer Ahmed Malik the definitions have been checked again and again by eminent experts in order to check its Authenticity keeping in mind that a single source is not enough in this regard. My sole aim was simplicity, uncomplicated approach and to clear those points where confusion arises normally.

This glossary includes all the important terms which are normally used in floras and plant guides which are used normally in herbarium.

Learning outcomes are ensured at every stage, language is age-appropriate and easy to understand by the Undergraduate Students. The illustrations are attractive, clear and relevant to the text and can also be used as a teaching tool at some stages specially when required. It will help to prepare students to study more complex diagrams carefully and critically at higher studies which means this glossary will work as a foundation to complex structures which come across as the study goes broad. This glossary gives the precise meanings/definitions of important terms used in different books of Plant Sciences; correct use of glossary can help students become independent learners.

Apart from it, this glossary provides comprehensive information and learning structures, the selection of content is based on relevance, importance and objectivity. The illustrations attract and retain the "user" interest and attention, especially at this level, the language is simple and it will reinforce and asses students learning.

Overall, both teachers and students will enjoy this glossary and will become active partners in the learning of Plant Sciences which is indeed the mother of emerging sciences.

# An Illustrated Glossary of Botanical Terminologies

## *(An Easy Approach to Plants Terms)*

## *(First Edition)*

2

# Glossary of Alphabet (A) Terminologies

**Abaxial epidermis of ovary:** The epidermal cell layer of the abaxial or outer surface of the ovary.

**Abaxial nucellar projection:** A portion of plant tissue that is the abaxial or lower portion of the nucleus overlying the vascular strands in the grass caryopsis.

**Abaxial petiole canal:** A petiole canal on the abaxial surface of a petiole.

**Abaxial protoderm:** The outermost layer of the shoot apical meristem which gives rise to the abaxial or lower leaf epidermis.

**Abaxial side of leaf primordium:** A portion of phyllome primordium tissue that is the abaxial or outer side of a leaf primordium and develops into the abaxial or lower leaf blade.

**Abaxial:** The lower surface of the leaf, away from axis.

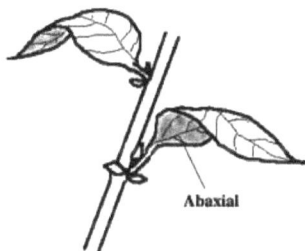

**Abruptly pinnate:** A compound leaf without a terminal leaflet.

**Abscission zone:** A portion of plant tissue that is part of a plant structure that parts a separation layer and a protective layer and is involved in the abscission of the structure.

**Acanceous:** Being prickly.

**Acantha:** A prickle or spine.

**Acanthocarpous:** Fruits are spiny.

**Acanthocladous:** Branches are spiny.

**Acarpous:** No pistil or carpillate whorl.

**Acaulescent:** (Stemless), the leaves and inflorescence rise from the ground appearing like that it have no stem.

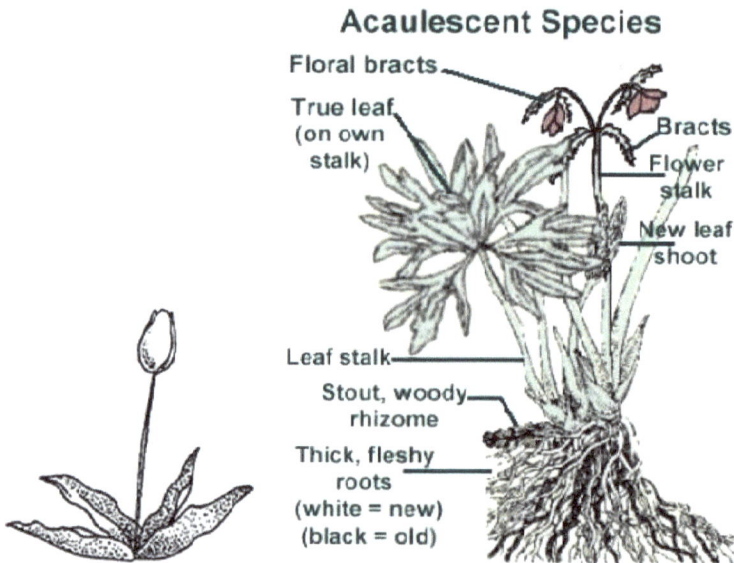

**Acaulescent Species**

**Accent:** Accent plants are utilized to bring attention to a specific plant characteristic and in turn draw the eye to the area of landscape they occupy.

Accent plants offer stunning foliage colour, growth habit and unique flowers or combination of all three.

**Accessory bud:** An embryonic shoot occurring above or to the side of an axillary bud.

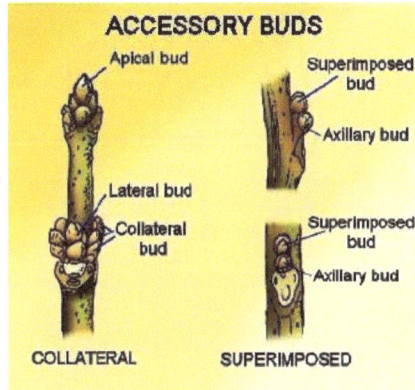

**Accessory paraclade:** A primary inflorescence branch borne in the same axil as an existing primary inflorescence branch.

**Accessory structures:** Parts of fruits that do not form, from the ovary.

**Accrescent:** Used for the calyx, when it is persistent and enlarges as the fruit grows and ripens, growing larger after anthesis.

**Accumbent:** Seeds in which the embryonic root is wrapped around and lies along the edges of the two cotyledons.

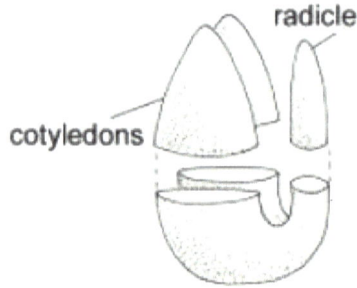

**Acephalous:** Used for flower styles without a well-developed stigma.

**Acerose:** Sharp, needle shape.

**Achene:** A small dry one seeded indehiscent fruit deriving from a one chambered ovary. *e.g.*

*Asteraceae.*

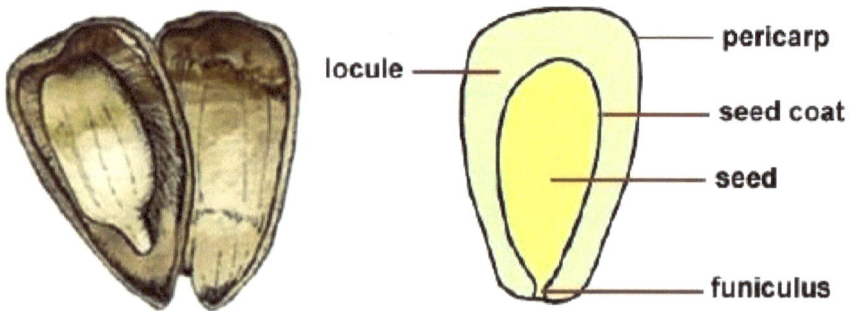

**Achlamydeous:** Flower without a perianth.

**Acicular:** Need like and slender shaped. *e.g.* some kind of foliage.

acicular

**Aciculate:** Fine lines usually randomly arranged, finely marked as pine pricks.

**Acme:** The period when the plant is at its maximum vigor.

**Acrandrous:** Used for moss species those have antheridia at the top of the stem.

**Acre:** A measure of land totaling of 43,560 square feet, a square feet is 208.75 feet on each side.

**Acrocarpous:** In mosses, bearing the sporophyte at the axis of the main shoot, produced at the end of the branch.

**Acrocaulous:** At the tip of the stem.

**Acrodramous:** When the veins run parallel to the leaf edge and fuse at the leaf tip.

**Acrogynous:** In liverworts, the female sex organs terminate the main shoot.

**Acropetal:** Toward apex, developing upward.

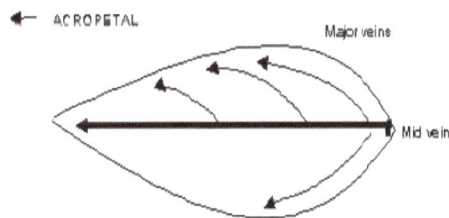

**Acroramous:** Terminal leaves, near apex of the branch.

**Acroscopic:** Facing apically.

**Actinodromous:** With three or more primary veins diverging radially from a single point at orabove the base of the blade and running towards the margin, reaching it or not.

**Basal-marginal -**
Midveins originate from the base
and secondaries reach the margin.

**Suprabasal-marginal -**
Midveins originate above the base,
within the leaf, and
secondaries reach the margin.

**Actinomorphic:** Parts of plants those are radially symmetrical in arrangement, as in regular flowers.

Cutting Planes

**Actinomorphic Flower**

**Actinostele:** A protostele having a xylem core in the form of radiating ribs, as view in transverse section.

**Aculeate:** Having a covering of prickles and needled like growth.

**Aculeolate:** Having spine like processes.

**Acuminate:** Tapering gradually to a long point.

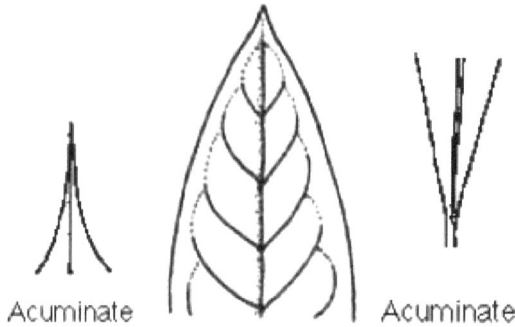

**Acutangular:** A stem that has several longitudinally running ridges with sharp edges.

**Acute:** Tapering to a sharp rather abrupt pointed apex with more or less straight sides along the tip.

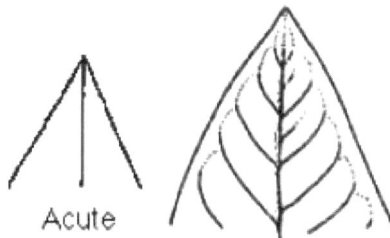

**Acutifolius:** With acute leaves.

**Acyclic:** With the floral parts arranged spirally rather than in whorls.

**Adaxial epidermis of ovary:** The epidermal cell layer of the adaxial or inner surface of the ovary.

**Adaxial nucellar projection:** A portion of plant tissue that is the adaxial or upper portion of the nucellus overlying the vascular strands in the grass caryopsis.

**Adaxial petiole canal:** A petiole canal on the adaxial surface of a petiole.

**Adaxial protoderm:** The outermost layer of the shoot apical meristem which gives rise to the adaxial/upper leaf epidermis.

**Adaxial side of leaf primordium:** A portion of phyllome primordium tissue that is the adaxialor inner side of the leaf primordium develops into the adaxial or upper leaf blade.

**Adaxial:** Located on the side facing towards the axis.

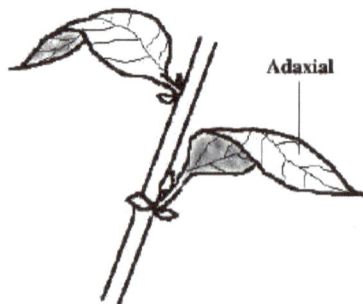

**Adelphous:** The androecium with the stamen filaments partly or completely fused togather.

**Aden:** A gland.

**Adenoid:** Gland like.

**Adenophore:** A stalk that supports a gland.

**Adenophorous:** Gland bearing.

**Adenophyllous:** Leaves with glands.

**Adherent:** With unlike parts of organs joined, but only superficially and without actual histological continuity.

**Adnate:** With unlike parts or organs integrally fused to one another with histological continuity.

**Adpressed:** Closely pressed together but not united.

**Adult vascular leaf:** A vascular leaf described by particular anatomical traits namely, wax and trichome distribution, presence or absence of epidermal cell types, cell wall shape and biochemistry.

**Aduncate:** Hooked.

**Adventitious bud:** A bud that arises at points on the plant other than the stem apex or leaf axil.

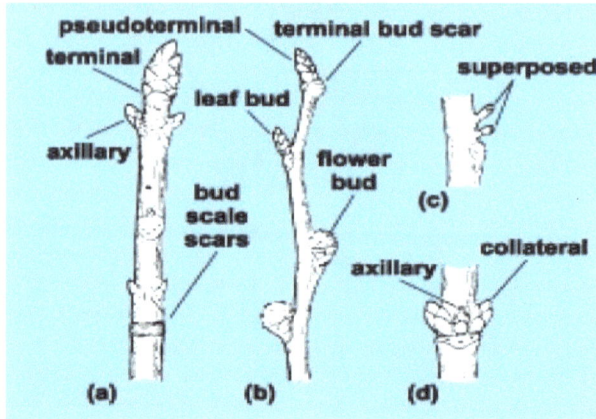

**Adventitious root apical meristem:** A root apical meristem that is part of an adventitious root.

**Adventitious root epidermis:** A portion of root epidermis that is part of an adventitious root.

**Adventitious root nodule:** Enlargement or swelling at position of dormant adventitious root primordium (on the stem), inhabited by nitrogen-fixing bacteria.

**Adventitious roots:** It develops from any part of plant other than radicle. In other words, adventitious roots do not arise from radicle.

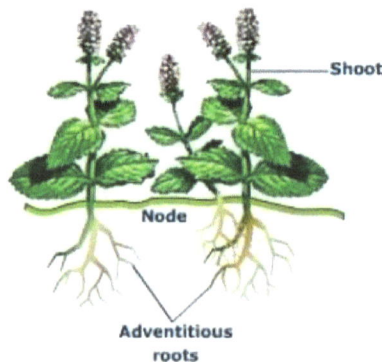

**Adventitious:** Occurring in unusual or unwanted locations such as roots on aerial stems or budson leaves.

**Aerate:** Loosening or puncturing the soil to increase water penetration.

**Aerenchyma:** A portion of parenchyma tissue containing particularly large intercellular spaces of schizogenous or lysigenous origin.

**Aerial stem:** Above the ground or water, in the air.

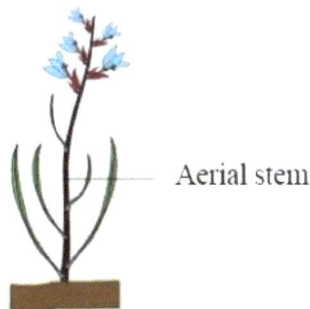

Aerial stem

**Aerial tuber axillary bud meristem:** A tuber axillary bud meristem that is part of an aerial tuber.

**Aerial tuber axillary shoot:** An axillary shoot that develops from an aerial tuber axillary bud.

**Aerial tuber axillary vegetative bud:** A tuber axillary vegetative bud that is part of an aerial tuber.

**Aerial tuber cortex:** A portion of tuber cortex that is part of an aerial tuber storage parenchyma.

**Aerial tuber epidermis:** A portion of tuber epidermis that is part of an aerial tuber.

**Aerial tuber interfascicular region:** A tuber inter fascicular region that is part of an aerial tuber storage parenchyma.

**Aerial tuber periderm:** A portion of tuber periderm that is part of an aerial tuber.

**Aerial tuber perimedullary zone:** A tuber perimedullary zone that is part of aerial tuber pith.

**Aerial tuber pith:** A portion of tuber pith that is part of an aerial tuber storage parenchyma.

**Aerial tuber storage parenchyma:** A portion of tuber storage parenchyma that is part of an aerial tuber.

**Aerial tuber:** A shoot axis tuber that develops above ground.

**Aerial:** An erect stem arising from a horizontal rhizome.

**Aerocaulous:** With aerial stems.

**Aerophyllous:** With aerial leaves.

**Aestival:** Appearing in summer.

**Aestivate:** To become dormant in summer.

**Aestivation:** The arrangement of sepals and petals in bud conditions is called aestivation.

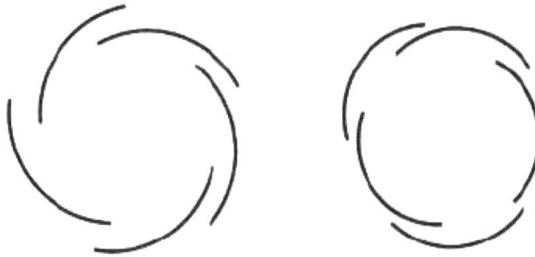

**Agamandrous:** Inflorescence with neuter flowers inside or above and staminate outside or below.

**Agamous:** Sexual organs abortive, without sex.

**Agglomerate:** Dense structures with varied angles of divergence.

**Agmohermaphroditic:** Inflorescence with neuter flowers inside or above and hermaphroditic outside or below.

**Aianthous:** With flowers appearing throughout the year.

**Air layering:** A specialized method of plant propagation accomplished by cutting into the bark of the plant to induce new roots to form.

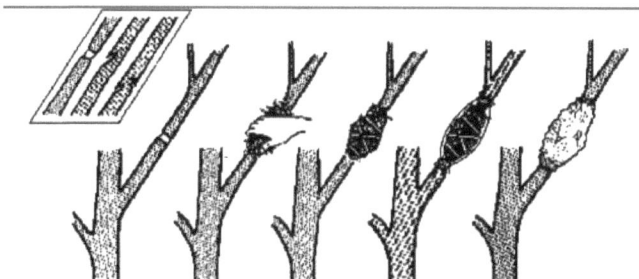

**Alar cell:** A plant cell located at the base of a non-vascular leaf adjacent to where the leaf attaches to the stem.

**Alate:** Having wing or wing like structures.

**Albuminous cell:** A parenchyma cell that is morphologically and physiologically associated with a sieve cell but does is not derived from the same initial cell as the sieve cell.

**Aleurone layer:** A portion of plant tissue that is the outermost layer of endosperm in a seed, its cells being characterized by presence of protein bodies containing seed storage proteins.

**Allagostemonous:** Having stamens attached to petal and torus alternately.

**Allautogamy:** Cross and self-fertilization in same plant.

**Allele:** alternate forms of a genetic locus. For example, at a locus determining eye colour, an individual might have the allele for blue eyes, brown.

**Allelopathy:** A characteristic of some plants according to which chemical are produced that inhibit the growth of other plants in the immediate vicinity.

**Allogamy:** When one plant pollinates another plant, cross pollination.

**Allopatric:** Occupying different geographic regions.

Allopatric Speciation

**Alternate:** Having structure in two rows or series so that the inner structure has its margins overlapped by a margin from each adjacent outer structure.

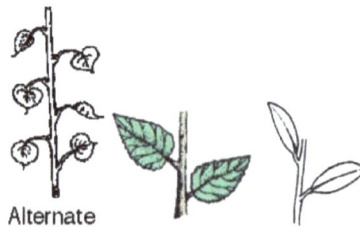

Alternate

**Alveolate:** Honey combed.

**Ammophilous:** Sand loving.

**Amphicarpous:** With fruits in two environments. *E.g.* aerial and subterranean.

**Amphiflorous:** Flowers and fruits above and below ground.

**Amphithecium:** The external cell layer of the developing sporangium of bryophytes.

**Amphithecium:** A portion of plant tissue that is the outer layer or layers of a sporangium earlyin sporangium development.

**Amplexicaul:** A sessile leaf that has its base completely surrounding the stem.

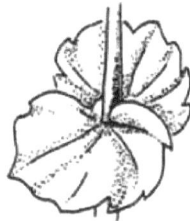

**Ampliate:** Dilated, enlarged.

**Anacrogynous:** In liverworts, female sex organs are produced by a lateral cell, thus the growth of main shoot is in determinate.

**Anadromous:** Having the first lobe or segment of a pinna arising basically in compound leaves.

**Anandrous:** Without stamens.

**Ananthous:** Without flowers.

**Anatomy:** It is the study of internal morphology or internal structure of plant and its organs.These characters are important for the study of correlated evolutionary sequences.

**Ancipital:** Two edged.

**Andradioecious:** Some plants with staminate flowers and some with perfect flowers.

**Andragamous:** Inflorescence with staminate flowers inside or above and neuter flowers outside or below.

**Androecium primordium:** A floral structure primordium that is committed to the development of an androecium.

**Androecium:** It is the male reproductive part of the flower. Stamens are collectively called as Androecium. Each stamen consists of two parts: 1. Filament. 2. Anther.

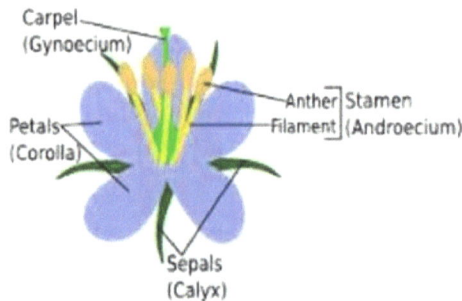

**Androgynecandrous:** Inflorescence with staminate flowers above or below pistilate, as in the spikes.

**Androgynous:** Monoecious and producing both types of sex organs together used for the inflorescence, when a spike has both staminate and pistilate flower, the pistilate flowers are normally at the base of spike.

**Androhermaphroditic:** Inflorescence with staminate flowers inside or above and hermaphroditic outside or below.

**Andromonoecious:** Plants with staminate and perfect flowers.

**Anemophlious:** Wind pollinated.

**Angled cuts:** When pruning branches it is recommended to make angled cuts, the cuts should be between 45 and 60 degrees to the plane of the branch.

**Angled:** Sided, as in the shape of the stems or fruits.

**Angular:** Having sharp angles or corners.

**Angustate:** Narrow.

**Anisocarpous:** With un equal carpels.

**Anisocotylous:** With unequal cotyledons.

**Anisolateral:** With unequal sides.

**Anisopetalous:** With unequal petals.

**Anisophyllous:** With unequal leaves.

**Anisosporous:** In dioecious bryophytes meiosis produces two small spores that develop into male gametophytes and two larger spores that develop into female gametophytes.

**Anisostylous:** With unequal styles.

**Annotinal:** Appearing yearly.

**Annuals:** The plants that complete their life cycle in one growing season are called annuals.New plants are produced by the seeds developed on these plants. *e.g.* Wheat (*Triticum aestivum*), Maize (*Zea mays*), Rice (*Oryza sativa*).

**Annular:** In the form of a ring.

**Annulus:** In mosses, cells with thick walls along the rim of the sporangium and were the peristome teeth are attached.

**Anterior:** On the front side away from the axis.

**Anther dehiscence zone:** A dehiscence zone that is part of an anther.

**Anther locule:** A sporangium locule that is part of an anther and is a cavity within a single pollen sac or two or more fused pollen sacs.

**Anther pore:** A plant anatomical space that is a pore at the apex of an anther.

**Anther primordium:** A floral structure primordium that is committed to the development of an anther.

**Anther septum:** A septum that is present in an anther dehiscence zone.

**Anther theca:** A collective plant organ structure that is part of an anther and consists of a pair of sporangia that dehisce down a common slit.

**Anther vascular system:** A phyllome vascular system that includes the totality of the portions of vascular tissue in their specific arrangement in a anther.

**Anther wall endothecium:** A microsporangium endothecium that part of an anther wall.

**Anther wall exothecium:** A microsporangium exothecium that is part of an anther wall.

**Anther wall inner secondary parietal cell layer:** An anther wall secondary parietal cell layer that is formed towards the inside after the cells of an anther wall primary parietal cell layer undergo a periclinal division.

**Anther wall middle layer:** A portion of ground tissue directly internal to an anther endothecium that develops from an anther wall parietal cell layer.

**Anther wall outer secondary parietal cell layer:** An anther wall secondary parietal cell layer formed towards the outside after an anther wall primary parietal cell layer undergoes a periclinal division.

**Anther wall primary parietal cell layer:** A portion of ground tissue that is part of an anther wall and has as parts a primary parietal cell and adjacent cells.

**Anther wall secondary parietal cell layer:** A portion of ground tissue that is part of an antherwall and develops from an anther wall primary parietal cell layer.

**Anther wall tapetum cell:** A microsporangium tapetum cell that is part of the anther wall tapetum.

**Anther wall tapetum:** A microsporangium tapetum that is part of an anther wall.

**Anther wall:** A microsporangium wall that is part of an anther.

**Anther:** The pollen-bearing portion of a stamen, the distal end of the stamen where pollen is produced, normally composed of two parts called anther-sacs and pollen-sacs.

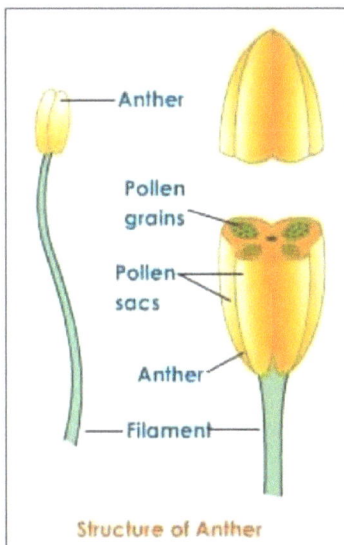

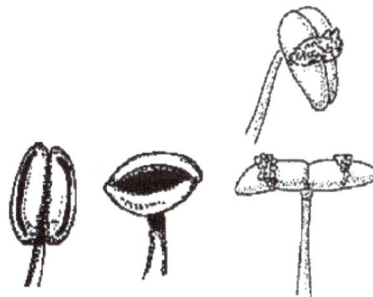

Structure of Anther

**Antheridiophore:** A specialized branch that bears antheridia.

**Antheridiophore:** A stalk that supports an antheridium head.

**Antheridium head:** A disk-shaped cardinal organ part that is the apical portion of an antheridiophore and bears antheridia.

**Antheridium jacket layer cell:** A plant cell that is part of an antheridium jacket layer.

**Antheridium jacket layer:** A portion of plant tissue that is a single layer of cells on the outside of an antheridium.

**Antheridium microgametophyte:** A microgametophyte that has as parts one or more antheridia.

**Antheridium sperm cell:** A plant sperm cell which develops from a spermatogenous cell and is located in an antheridium.

**Antheridium stalk:** A stalk that is the basal part of an antheridium.

**Antheridium:** The male sex organ producing the sperm.

**Antheridium:** A multicellular plant gametangium that produces antheridium sperm cells and has as parts an antheridium jacket layer and an antheridium stalk.

**Anthesis:** The period when the flower is fully open and functional, ends when the stigma or stamens with her.

**Anthocarpous:** Having a body of combined floral and fruit parts, as in multiple fruits.

**Anthology:** Study of flower.

**Anthotaxis:** Arrangment of sporophylls.

**Anthropophily:** Pollinated by man.

**Antipetalous:** When the stamens are the same number as the corolla segments and oppositely arranged the corolla segments.

**Antipodal cell:** Cell, commonly three in numbers as in the eight-nucleate embryo sac, located atthe other end of the embryo sac from the egg cell.

**Antiraphe:** A cardinal organ part that is the part of a plant ovule on the opposite side from araphe.

**Antisepalous:** When the stamens are the same number as the calyx segments and oppositely arranged the calyx segments.

**Antitropous:** With radicle pointing away from hilum.

**Antrorse:** Pointing forward or upward.

**Aperturate:** With one or more openings or apertures.

**Apetalous:** No petals or corolla, a flower without petals.

**Apex:** The tip of a plant part.

**Aphyllopodic:** Without blade bearing leaves at base of plant.

**Aphyllous:** Without leaves, no whorls of leaves.

**Apical hook:** Hook-like structure which develops at the apical part of the hypocotyl in dark-grown seedlings in dicots.

**Apical meristem:** A maximal portion of meristem tissue located at an apex of a shoot system or root system.

**Apical placentation:** The attachment of ovules is at the apex of the ovary, one locule, no septa.

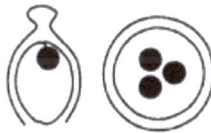

**Apical:** Growth region at the apex of the structure.

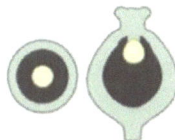

**Apiculate:** Ending in an abrupt slender tip which is not stiff.

**Apocarpous:** With separate carpels.

Apocarpous
spiralled

Apocarpous
whorled

**Apogamy:** Producing sporophytes from a gametophyte without fertilization.

**Apopetalous:** With separate petals.

**Apopyhsis:** Exposed outer surface of either an ovuliferous scale or megasporophyll as seen when the cone is closed.

**Aposepalous:** With separate sepals.

**Apospory:** Producing gametophytes directly from a sporophyte without producing spores.

**Apostemonous:** With separate stamens.

**Applanate:** Flat, without vertical curves or bends.

**Appressed:** Lying flat against or nearly parallel to, as leaves on a stem or hairs on a leaf.

**Arachnoid:** Having a cobwebby appearance with entangled hairs.

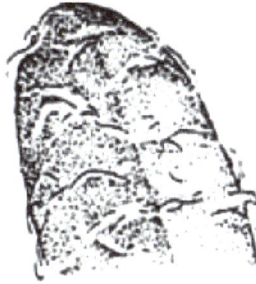

**Arborescent:** Growing into a tree like habit, normally with a single woody stem.

**Arboretum:** A garden with a large collection of trees and shrubs cultivated for scientific purposes.

**Archegoniophore:** A specialized branch that bears the archegonia.

**Archegonium central cell:** A plant cell that is the larger, distal cell arising from the first division of an archegonium initial cell.

**Archegonium egg cell:** A plant egg cell that develops from the archegonium central cell and is located in the venter.

**Archegonium head:** An umbrella-shaped cardinal organ part that is the apical portion of anarchegoniophore and bears archegonia.

**Archegonium initial cell:** An initial cell that divides asymmetrically to form an archegonium central cell and a smaller archegonium neck canal cell.

**Archegonium megagametophyte:** A mega gametophyte that has as parts one or more archegonia.

**Archegonium neck canal cell:** A plant cell that is one of the axial rows of cells in an immature archegonium neck.

**Archegonium neck canal:** A canal in the center of an archegonium neck.

**Archegonium neck:** A cardinal organ part that is the elongated, apical part of an archegonium.

**Archegonium stalk:** A stalk that the basal part of an archegonium.

**Archegonium:** A multicellular plant gametangium that develops from an archegonium initial cell and has as parts a venter and an archegonium neck.

**Archesporial cell:** A plant cell that divides to gives rise to a sporocyte and is part of a sporangium.

**Archesporium:** A portion of plant tissue that is the internal part of a sporangium, bounded by the sporangium wall, and has as parts archesporial cells.

**Arcuate:** Curved like a crescent, can be downward or upward.

ARCUATE

**Areolate:** Divided into many angular or squarish spaces.

**Areole bud:** An axillary vegetative bud that is not elongated, in which the vascular leaves develop as spine leaves.

**Areole:** A raised area on a*cactus* from which spines develop.

**Areoles:** The spaces formed by a vein network.

**Arhizous:** Without roots, no whorls of roots.

**Aril:** An outgrowth from the stem forming a fleshy covering of the seed or only rudimentary at base of fleshy seed.

**Arillode:** An arilloid that is an elaborate outgrowth of a seed at the micropylar end.

**Arilloid:** A cardinal part of multi-tissue plant structure that is a fleshy outgrowth of a seed.

**Aristate:** Ending in a stiff, bristle like point, typically at the apex.

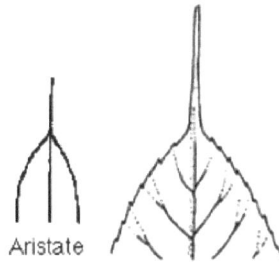

Aristate

**Armed:** Provided with prickles, spines or thorns.

**Articulate:** Having a joint as in leaves, leaflets or stems or having a swollen area often discolored at the point of branching of the stem.

**Articulated laticifer cell:** A laticifer cell that is joined longitudinally to other articulated laticifer cells to form a tube.

**Articulated laticifer:** A portion of secretory tissue that has as parts articulated laticifer cells.

**Ascending:** Growing obliquely uprightly, in an upward direction, heading in the direction of the top.

**Asepalous:** No sepals or calyx.

**Asperous:** Having a rough surface.

**Assurgent:** Directed upward or forward.

**Astemonous:** No stamens or androecium.

**Asymmetric:** Without regularity in any dimension.

**Asymmetrical:** Not divided into like and or equal parts.

**Attenuate:** Tapering gradually to a narrow end.

**Auriculate:** Lobe rounded, sinus depth variable, outer margin concave, inner convex or straight.

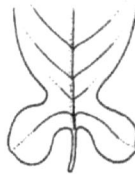

**Auriculiform:** Usually obovate with two small rounded, basal lobes.

**Austral:** Southern.

**Autogamy:** Self-pollination, when the flowers of the same plant pollinate flowers on the same plant or themselves.

**Autoicous:** Produces male and female sex organs on the same plant but on separate inflorescences.

**Autophilous:** Self-pollinated.

**Autumnal:** Appearing in autumn.

**Awl-shaped leaf:** Subulate, narrow, stiff, flat, sharp pointed leaf.

**Awn:** A cardinal organ part that is a slender, more or less straight and stiff, fine-pointed, terminal or subterminal appendage of a glume, lemma, or palea.

**Axial secondary xylem parenchyma cell:** A secondary xylem parenchyma cell that is part of aportion of axial secondary xylem parenchyma.

**Axial secondary xylem parenchyma:** A portion of secondary xylem parenchyma that is part of an axial system and has as parts axial secondary xylem parenchyma cells.

**Axial system:** A portion of secondary vascular tissue that has as parts cells derived from fusiform initials and oriented with their longest diameter parallel to a plant axis.

**Axil:** The upper angle formed between two structures or organs, such as a leaf and the stem from which it grows.

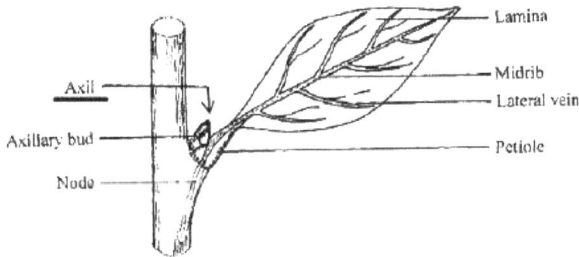

**Axile placentation:** Ovules are attached to an axis derived from the connate margins of the component carpals, such that an ovary is divided into two or more locules by septa, the ovules are borne along the central axis, and can only be found in a syncarpous gynoecium.

**Axile:** With the placentae along the central axis in a compound ovary with septa.

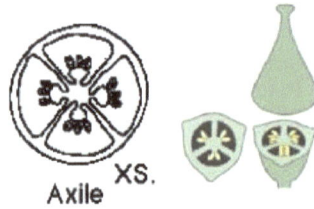

Axile XS.

**Axillary bud meristem:** A shoot meristem formed in an axil.

**Axillary bud:** A bud that develops from an axillary bud meristem.

**Axillary flower bud:** An axillary bud that develops into a flower.

**Axillary hair basal cell:** A trichome cell that is part of a base of an axillary hair.

**Axillary hair base:** A portion of plant tissue that is the basal part of an axillary hair, below the axillary hair terminal cell.

**Axillary hair terminal cell:** A trichome cell that is the long terminal cell of an axillary hair.

**Axillary hair:** A multicellular trichome that has as parts a long terminal cell atop a basal stalk and grows in a leaf axil.

**Axillary inflorescence bud:** An axillary reproductive bud that develops into an inflorescence.

**Axillary reproductive bud:** An axillary bud that develops into a reproductive shoot system.

**Axillary shoot system:** A shoot-borne shoot system that develops from an axillary bud.

**Axillary strobilus bud:** An axillary reproductive bud that develops into a strobilus.

**Axillary vegetative bud:** An axillary bud that develops into a shoot system that has only vegetative organs as organ parts.

**Axillary:** An embryonic shoot which lies at the junction of the stem and petiole of a plant.

Axillary

**Axis:** The main stem.

# Glossary of Alphabet (B) Terminologies

**Baccate:** Juicy and succulent.

**Balsamiferous:** Sticky and aromatic, like balsam.

**Banded:**Transverse stripes of one color crossing another.

**Banner petal:** A petal that is the top-most petal of a papilionaceous corolla.

**Banner:** The upper petal of a pea flower.

**Barbed:** With short, rigid reflexed bristles or processes.

**Barbellate:** Minutely barbed.

**Bare root:** Plants offered for sale which have had all of the soil removed from their roots.

**Bark:** The outer layer of woody plants; cork, phloem, and vascular cambium.

**Bark:** All tissues outside the vascular cambium or the xylem; in older trees may be divided intodead outer bark and living inner bark, which consists of secondary phloem.

**Basal axillary shoot system:** An axillary shoot system that is part of a stem base.

**Basal endosperm transfer cell:** A transfer cell that is part of a basal endosperm transfer layer.

**Basal endosperm transfer layer:** A portion of plant tissue that is part of an endosperm and is composed of basal endosperm transfer cells.

**Basal placentation:** Attachment of ovules to the bottom of the overy. One locule, no septa.

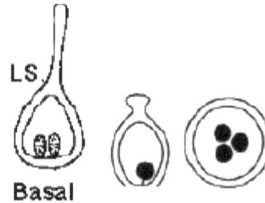

**Basal root primordium:** A root primordium that is committed to the development of a basal root.

**Basal root:** A root that arises from a part of the hypocotyl.

**Basal:** Growth region at the base of a blade as in grasses.

**Basicaulous:** Near base of stem.

**Basifixed:** Attached by the base.

**Basipetal:** Developing downward, toward base.

**Basipetiolar:** At the base of the petiole.

**Basiramous:** Leaves on lower part of branch.

**Basiscopic:** Facing basally.

**Bast bundles:** Bundles of thick-walled cells parallel to the midrib, as in isoetes.

**Beak:** A firm, pointed terminal appendage, normally the slender elongated end of a fruit, typically a persistent style-base.

**Bearded or barbate:** With long trichomes usually in a tuft, line or zone.

**Bedding plant:** Plants, nursery grown and suitable for growing in beds. Quick, colorful flowers.

**Berry:** A fleshy, indehiscent fruit in which the seeds are not encased in a stone and are typically more than one.

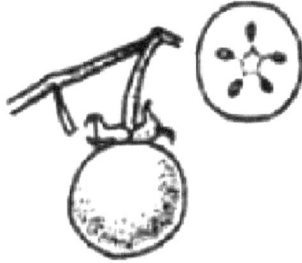

**Bicarpellate:** Two carpelled.

**Bicrenate:** With smaller rounded teeth on larger rounded teeth.

**Bidentate:** Two-toothed.

**Biduous:** Lasting two days.

**Biennials:** The plants that complete their life cycle in two growing seasons are called biennials. During first year, leaves are produced while in second year flowers and seeds are produced. *E.g.* Radish (*Raphanus sativus*), Carrot (*Daucus carota*).

**Bifarious:** In two vertical rows.

**Bifid:** Cut or divided in to two lobes or parts.

**Bifoliate:** Two-leaved.

**Bifolrous:** Flowering in autumn as well as in spring.

**Bifurcate:** Divided in to two forks or branches.

**Bilabiate:** Two-lipped.

**Bilocular:** Tow-locular.

**Bimestrial:** Occurring every two months.

**Binate:** Twinned.

**Bipinnate:** Twice pinnately compound, compound leaf structures with a feather-liked formation of leaflets arranged in pairs, with each leaflet also pinnately divided in to pairs.

**Bipinnatifid:** Two times pinnately cleft.

**Biserrate:** With sharply cut teeth on the margins of larger sharply cut teeth.

**Bisexual:** Having both stamens and inflated.

**Blade:** The expanded terminal portion of leaf, petal or another structure, *i.e.* that portion of the leaf that does not include the stalk.

**Blastocarpous:** Germination of seeds while with in pericarp, as in Rhizophora.

**Bloom:** A white, powder like coating sometimes found on a leaf or stem surface, the waxy coating that cover some plants.

**Blotched:** The color disposed in broad, irregular blotches.

**Body cell:** A spermatogenous cell of the microgametophyte that divides to produce a vegetative cell and the pollen sperm cell.

**Bole:** The trunk or stem of a tree.

**Bolting:** Vegetables which quickly go to flower rather than producing the food crop. Usually caused by late planting and too warm temperatures.

**Bone meal:** Bone meal is an excellent natural fertilizer for a wide variety of plants including flowers, bulbs, roses and vegetables.

**Boreal:** Northern.

**Botanical name:** The scientific name of a plant usually composed of two words, the genus and the species.

**Botuliform:** Sausage shaped.

**Brachycyte:** A plant cell that develops from a cell in a protonema and has a thick cell wall.

**Brackish:** A mixture of salt and fresh water, somewhat saline.

**Bract adaxial epidermis:** A portion of bract epidermis that covers the adaxial/upper surface of abract.

**Bract anlagen:** A phyllome anlagen that will give rise to a bract primordium and is part of aperipheral zone of a reproductive shoot apical meristem.

**Bract apex:** A phyllome apex that is part of a bract.

**Bract axil:** An axil that is the space between a shoot axis and a bract that branches from the shoot axis.

**Bract base:** A phyllome base that is part of a bract.

**Bract epidermis:** A portion of phyllome epidermis that is part of a bract.

**Bract margin:** A phyllome margin that is part of a bract.

**Bract primordium:** A phyllome primordium that is committed to the development of a flower bract.

**Bract stomatal complex:** A phyllome stomatal complex that is part of a bract.

**Bract tip:** A phyllome tip that is part of a bract apex.

**Bract trichome:** A phyllome trichome that is part of a bract epidermis.

**Bract:** Leaf is present below the flower, modified scale like leaves usually growing just below a flower or flower cluster.Often confused with petals or the flower itself.Bracteole and laminar. May be localized or found over entire structure.

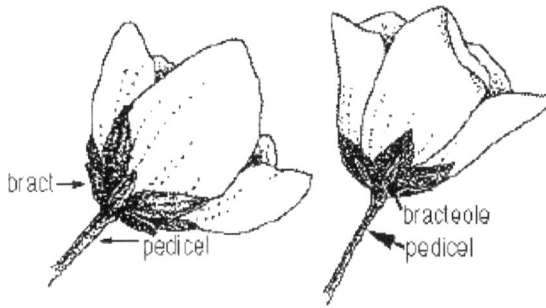

**Bracteole:** A small leaf or leaves borne singly or in pairs on the pedicel, the prophyll of the flower-shoot.

**Branch axil:** An axil that is the space between a shoot axis and a branch that branches from the shoot axis.

**Branch internode differentiation zone:** A shoot internode differentiation zone that is part of a branch internode.

**Branch internode elongation zone:** A shoot axis internode elongation zone that is part of a branch internode.

**Branch internode:** A shoot inter node that is part of a branch.

**Branch node:** A shoot node that is part of a branch.

**Branch procambium:** A portion of shoot axis procambium gives rise to the primary vascular tissue of a branch.

**Branch stele:** A shoot axis stele that is part of a branch.

**Branch tendril:** A branch that is slender and coiling.

**Branch trichome:** A shoot axis trichome that is part of a branch epidermis.

**Branch:** A shoot axis that develops from an axillary bud meristem or from equal divisions of a meristematic apical cell.

**Branching:** Dividing in to multiple smaller segments.

**Breeding**: The intentional development of new forms or varieties of plants or animals by crossing, hybridization, and selection of offspring for desirable characteristics.

**Bristle:** A stiff hair, usually erect or curving away from its attachment point.

**Brochidodromous:** With a single primary vein, the secondary veins not terminating at the margin but joined together in a series of prominent upward arches or marginal loops on each side of the primary vein.

**Brunescent:** Brownish.

**Bud scale leaf:** A scale leaf that surrounds a dormant or perennating bud.

**Bud:** It is the compact structure formed on stem by short internodes and smaller overlapped leaves or immature shoot.

**Bulb:** These are reduced underground stem having disc at the base from which roots arise and itis covered by thick fleshy leaves having stored food. *e.g. Allium cepa* (onion)*, Allium sativum* (Garlic).

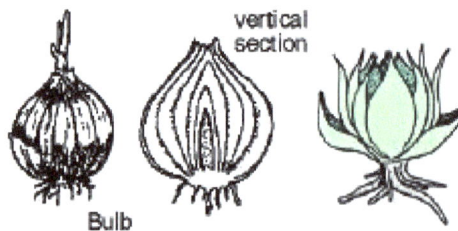

Bulb

**Bulblet:** A small, bud like vegetative propagule produced on the leaves of some ferns.

**Bulliform cell:** An enlarged more or less thin-walled leaf pavement cell, present, with other similar cells, in longitudinal rows in leaves of monocots.

**Bundle scar:** A small mark on the leaf scar indicating a point where a vein from the leaf was once connected with the stem.

**Bundle sheath cell:** A ground tissue cell that is part of a bundle sheath.

**Bundle sheath chlorenchyma cell:** A chlorenchyma cell that is part of a bundle sheath.

**Bundle sheath extension:** A strip of ground tissue presents along the leaf veins and extending from the bundle sheath to the epidermis. It may be present on one or both sides of the vein and may consist of parenchyma or sclerenchyma.

**Bundle sheath:** A layer or layers of cells surrounding the vascular bundles of leaves.It may consist of parenchyma or sclerenchyma.

**Bur:** A prickly or spiny seed or fruit.

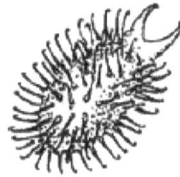

**Burl:** A woody swelling where the stem joins the roots.

# Glossary of Alphabet (C) Terminologies

**Caducous:** Falling away early, falling off very early as compared to similar structures in other plants.

**Caerulescent:** Bluish.

**Caespitose:** Growing in tufts.

**Calathiform:** Basket or cup shaped.

**Calciphilous:** Lime loving.

**Callus parenchyma cell:** A parenchyma cell that is part of a plant callus.

**Callus:** A hardened or thickened area at the point of attachment.

**Calyptra calyx:** A fused calyx that is composed of fused sepals.

**Calyptra corolla:** A fused corolla that is composed of fused petals.

**Calyx:** The whorl of sepals at the base of a flower, the outer whorl of the perianth.

**Cambial initial cell:** An initial cell that is part of the vascular or cork cambium and by periclinal divisions produces cells to the outside or inside of the cambial axis.

**Cambial zone:** A cardinal organ part that is part of a plant axis and has as parts a cambium and adjacent cells.

**Cambium:** The thin membrane located just beneath the bark of a plant.

**Cambium:** A lateral meristem that has as parts a single layer of cambial initial cells and their derivatives, arranged orderly in radial files.

**Campanulate:** Bell shaped.

**Canal:** A plant anatomical space that is either a groove on the surface or a tube in the interior of a plant structure.

**Canes:** Canes are woody stems of plants like roses and blackberries which produce their maximum amount of flowering or fruiting within a one or two year period. These canes are usually pruned severely or removed after their productive stage has passed to allow for new growth to take their places.

**Canescent:** With gray or white short hairs, often having a hairy appearance.

**Canopy:** The canopy is the highest level of branches and foliage providing shade below. Many plants which flourish with partial or dappled shade are well suited for growing under a canopy. Certain large trees will produce a canopy which will shade their lower limbs causing these branches to diminish in vigor. After these limbs are removed, the possibility could arise for new plantings. Referred to as understory plantings, suitable candidates will form a new architectural level to a landscape setting.

**Cantharophilous:** Beetle pollinated.

**Capillary:** Slender and hair like.

**Capitate:** In a globular or head-shaped cluster.

**Capitulescence:** A special term used in *Asteraceae* to describe a group of associated heads also called capitula; it is analogous to an inflorescence.

**Capitulum:** The flowers are arranged into a head composed of many separate unstalked flowers; the single flowers are called florets and are packed close together, the typical arrangement of flowers in the *Asteraceae*.

**Capreolate:** With tendrils.

**Capsule:** A dry generally many seeded fruit divided into two or more seed compartments that dehisces or splits open longitudinally with the line of dehiscence either through the locule or through the septa or less commonly through pores or around the circumference.

**Cardinal organ part:** A cardinal part of multi-tissue plant structure that is a proper part of a plant organ and includes portions of plant tissue of at least two different types.

**Cardinal part of multi-tissue plant structure:** A plant structure that is a proper part of a multi-tissue plant structure and includes portions of plant tissues of at least two different types.

**Carinate:** Keeled with one or more longitudinal ridges.

**Carpel anlagen:** A phyllome anlagen that will give rise to a carpel primordium and is part of a peripheral zone of a flower meristem.

**Carpel epidermis:** A portion of phyllome epidermis that is part of a carpel.

**Carpel margin:** The margin of a carpel.

**Carpel primordium:** A phyllome primordium that develops from a carpel anlagen and is committed to the development of a carpel.

**Carpel stomatal complex:** A phyllome stomatal complex that is part of a carpel.

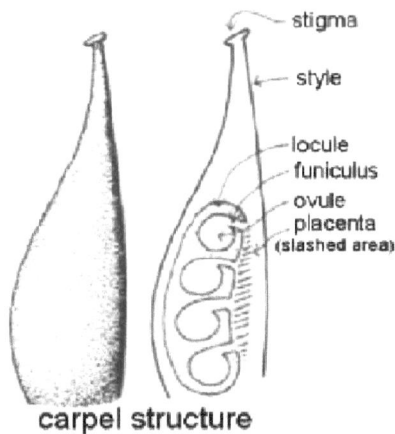

carpel structure

**Carpel trichome:** A phyllome trichome that is part of a bract epidermis.

**Carpel vascular system:** A phyllome vascular system that includes the totality of the portions of vascular tissue in their specific arrangement in a carpel.

**Carpel:** The ovule-producing reproductive organ of a flower consisting of the stigma, style and ovary.

**Carpel:** A megasporophyll, almost always at the center of a flower, its margins more or less used together or with other carpels to enclose the ovule.

**Caruncle:** An arilloid that is an outgrowth of a seed next to the micropyle.

**Caryopsis hull:** A collective phyllome structure that encloses a fruit of the Poaceae and develops from a dried lemma and palea.

**Caryopsis:** The grain or fruit of grasses.

**Castaneous:** Dark reddish brown.

**Catkin:** A spike like often pendulous, inflorescence of petalless unisexual flowers, either staminate or pistillate.

**Caudate:** Bearing a tail or slender tail like appendage.

**Caudex:** The hard base produced by herbaceous perennials used to over-winter the plant, the persistent, often woody base of an otherwise annual herbaceous stem.

**Caulescent:** With a well-developed distinctive stem above ground.

**Cauline axillary shoot:** An axillary branch that forms from a leaf above the very base of the shoot.

**Cauline leaf:** A vascular leaf borne on the stem.

**Cauline paraclade:** A primary inflorescence branch borne on the elongated first order inflorescence axis.

**Cauline:** Attached to or referring to the stem as opposed to basal often used to describe leaf position.

**Caulonema cell:** A chlorenchyma cell that is part of a caulonema and has cross walls of adjacent cells that are oblique to the protonema axis.

**Caulonema meristematic apical cell:** A protonema meristematic apical cell that is part of a caulonema.

**Caulonema:** A portion of protonema tissue that consists of only caulonema cells.

**Cellular endosperm:** Endosperm in which the first karyokinesis event is accompanied by cytokinesis.

**Centarch protoxylem:** A portion of protoxylem tissue in which the maturation of primary xylem of the shoot system progresses centrifugally with the oldest elements (protoxylem) in the center of the axis.

**Central endosperm:** A portion of plant tissue that is the central region of an endosperm, composed of cells that are significantly larger than those at the periphery, especially the aleurone and sub-aleurone layers.

**Central placentation:** Ovules attached to a free standing central column in a syncarpous, unilocular ovary.

**Central root cap of lateral root:** A central root cap that is part of a root cap of lateral root.

**Central root cap of primary root:** A central root cap that is part of a root cap of primary root.

**Central root cap:** A portion of root parenchyma tissue that is the central part of a root cap in which the cells are arranged in longitudinal files.

**Central spike of ear inflorescence:** A first order inflorescence axis that is the central axis of a near inflorescence.

**Central spike of tassel inflorescence:** A first order inflorescence axis that is the central axis of a tassel inflorescence.

**Central strand:** A portion of plant tissue that is an axial strand in the center of a gametophores axis or seta and has as part a hydrome or a leptome.

**Central zone of the leaf lamina:** Region that includes the widest or majority portion of leaf lamina and that does not include leaf apex and leaf lamina base.

**Central zone of the petiole:** Region of the petiole that does not include petiole distal end and petiole proximal end.

**Central zone:** An area of densely packed cells in the shoot apex that divide infrequently.

**Ceonosorus:** A collective plant organ structure on the surface of a vascular leaf that has as parts two or more fused sori.

**Cernuous:** Nodding, drooping.

**Cespitose:** Forming dense tufts or cushion like growth, normally applied to small plants typically growing into mats tufts or clumps.

**Chaff:** Thin scales or bracts subtending individual flowers in many species of the *Asteracea.*

**Chalaza:** A portion of plant tissue that is the region in a plant ovule where the integuments and the nucelli join with the funicule.

**Chalazal cyst:** A portion of plant tissue that is the part of the endosperm nearest the chalazal region of the ovule and forming a cyst-like structure.

**Chambered pith:** A form of pith in which the parenchyma collapses or is torn during development, leaving the sclerenchyma plates to alternate with hollow zones.

**Chaparral:** An area characterized by dense, leathery-leaved, evergreen shrubs.

**Chiropterophilous:** Bat pollinated.

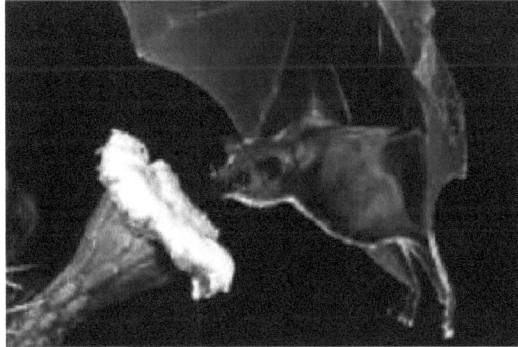

**Chlorenchyma cell:** A parenchyma cell containing chloroplasts; a component of leaf mesophyll and other green parenchyma tissue.

**Chlorenchyma:** Chloroplast-containing parenchyma tissue.

**Chloronema cell:** A chlorenchyma cell that is part of a chloronema and has cross walls of adjacent cells those are perpendicular to the protonema axis.

**Chloronema meristematic apical cell:** A protonema meristematic apical cell that is part of a chloroncma.

**Chloronema:** A portion of protonema tissue that consists of only chloronema cells.

**Chlorophyllous:** Having chlorophyll.

**Chlorotic:** Lacking chlorophyll.

**Chromosome**: The structures in the eukaryotic nucleus and in the prokaryotic cell that carries most of the DNA.

**Cigar leaf lamina abaxial epidermis:** A portion of leaf lamina abaxial epidermis that covers the abaxial surface of a cigar leaf.

**Cigar leaf:** A leaf near the apex of a stem that is still rolled into a cylinder.

**Cilia:** Marginal hairs.

**Ciliate:** With a fringe of marginal hairs, with a row of fine hairs situated along the marginof a structure such as a leaf.

**Cinereous:** Ash colored; light gray due to a covering of short hairs.

**Circumboreal:** Distributed around the globe at northern latitudes.

**Circumscissile:** A type of fruit dehiscences were the top of the fruit falls away like a lid or covering, dehiscing along a transverse circular line around the fruit or anther, so that the top separates or falls off like a lid.

**Cismontane:** Referring to the ocean-facing side as opposed to the desert-facing side of the mountains.

**Citreous:** Lemon yellow.

**Cladautoicous:** Male and female inflorescences on separate branches of the same plant.

**Cladode:** A flattened stem that performs the function of a leaf performing the process of photosynthesis, many plants with cladodes have no true leaves at all and if they do the leaves are miniscule and short lived. An example is the pad of the *Opuntia* cactus.

**Cladode:** A shoot axis that is flattened and expanded.

**Cladophyll:** A flattened stem that is leaf-like and green used for photosynthesis, normally plants have no or greatly reduced leaves.

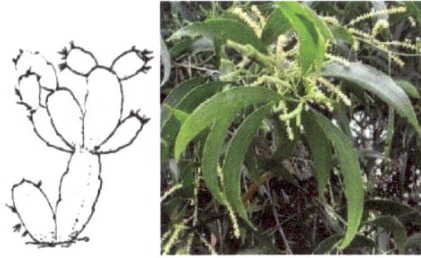

**Clasping:** Having the lower edges of a leaf blade partly surrounding the stem.

**Clavate:** Club-shaped, gradually thickened or widened toward the apex.

**Claw:** A noticeably narrowed and attenuate organ base, typically a petal.

**Cleft:** Deeply cut usually more than one-half the distance from the margin to the midrib or base.

**Cleistogamous:** Self-pollination of a flower that does not open.

**Climbing:** Typically long stems, that clings to other objects.

**Cold Hardiness:** Cold hardiness refers to a plant's ability to survive near-freezing and sub freezing temperatures.

**Coleoptile:** Protective sheathe on SAM.

**Coleoptile:** A sheath-like plant organ that surrounds the plumule of a plant embryo or the emerging shoot apex of a seedling.

**Coleorhiza:** Protecting layer of a seed.

**Coleorhiza:** A sheath-like plant organ that surrounds the radicle of a plant embryo or the emerging root of a seedling.

**Collar:** In grasses the outer side of the leaf at the junction of the sheath and blade.

**Collective leaf structure:** A collective phyllome structure composed of two or more leaves.

**Collective organ part structure:** A collective plant structure composed of two or more cardinal organ parts that are part of adjacent plant organs and any associated portions of plant tissue.

**Collective phyllome structure:** A collective plant organ structure that consists of two or more phyllomes originating from the same node or from one or more adjacent nodes with compressed shoot internodes.

**Collective plant organ structure:** A collective plant structure that is a proper part of a whole plant and is composed of two or more adjacent plant organs and the associated portions of plant tissue.

**Collective plant structure:** A plant structure that is a proper part of a whole plant and includes two or more adjacent plant organs or adjacent cardinal organ parts along with any associated portions of plant tissue.

**Collective tepal structure:** A perianth consisting of one or more tepals.

**Collenchyma cell:** An elongated plant cell with unevenly thickened non-lignified primary walls that is alive at maturity.

**Collenchyma:** Living tissue composed of more or less elongated cells with thick non-lignified primary cell walls.

**Columella root cap cell:** Cell that constitutes the central part of the root cap, arranged in longitudinal files.

**Columella root cap initial cell:** A root initial cell that produces columella cells in the root cap.In direct contact with quiescent center.

**Columella:** A portion of plant tissue that forms the central axis of a plant structure such as a fruit or moss capsule.

**Coma:** A tuft of hairs, often at the tip of seeds.

**Coma:** A portion of seed coat epidermis that consists of multiple seed trichomes at the micropyle.

**Companion cell:** A parenchyma cell that is adjacent to a sieve tube element and arises from the same phloem mother cell as the sieve tube element.

**Complete fertilizer:** A plant food which contains all three of the primary elements nitrogen, phosphorus and potassium.

**Complete:** Describing flowers that contain petals, sepals, pistils and stamens.

**Compost:** An organic soil amendment resulting from the decomposition of organic matter.

**Compound leaf:** Made up of two or more similar parts, as in a leaf which has leaflets.

**Compound leaf:** A leaf having two or more distinct leaflets that are evident as such from early in development.

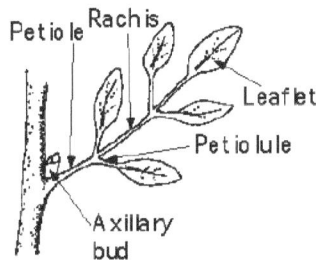

**Compound umbel:** An umbel where each stalk of the main umbel produces another smaller umbel of flowers.

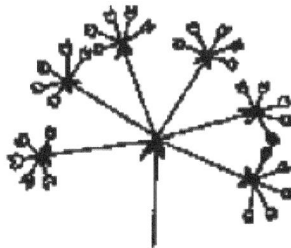

**Compression wood:** A portion of reaction wood found on the lower side of a shoot axis that isor was angled away from vertical and that has as parts heavily lignified tracheids with a S2 layer that contains more lignin and has a larger (more horizontal) microfibril angle.

**Concolor:** Uniform color.

**Confluent:** running together or blending of one part into another.

**Conifer:** A cone bearing tree with tiny needle like leaves.

**Connate:** When the same parts of a flower are fused to each other, petals in a gamopetalous flower, describing similar structures that are joined or grown together.

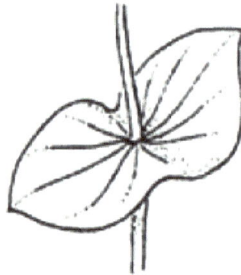

**Connective:** The part of the stamen joining the anther cells.

**Connective:** The part of the stamen that connects the microsporangia/pollen sacs.

**Connivent:** Converging, but not actually fused or united.

**Conspecific:** Same species.

**Contracted:** Narrowed or shortened as opposed to open or spreading.

**Convolute:** Rolled up longitudinally, with one edge inside the other and the upper surface on the inside.

**Cordate:** Heart-shaped, stem attaches to cleft.

**Coriaceouse:** Tough or leathery texture.

**Cork cambium:** A cambium that is part of a periderm and produces phellem and phelloderm.

**Corm:** A compact, upright orientated stem that is bulb-like with hard or fleshy texture and normally covered with papery, thin dry leaves. Most often produced under the soil surface.

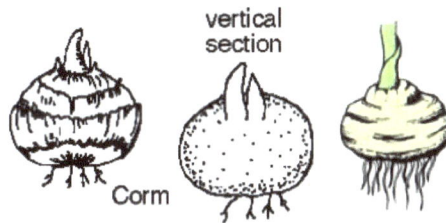

**Corneous:** Horny.

**Corniculate:** Having little horns or horn like appendages.

**Cornute:** Horned.

**Corolla spur:** A collective organ part structure that is a slender, hollow extension of a corolla and has as parts segments of two more fused petals.

**Corolla:** The whorl of petals of a flower, the inner whorl of the perianth between the calyx and the stamens, a collective term for the petals of a flower.

**Corolla:** A collective phyllome structure that is composed of one or more petals comprising the inner whorl of non-reproductive floral organs and surrounds the androecium and the gynoecium.

**Corona:** An additional structure between the petals and the stamens.

**Coroniform:** Crown-shaped.

**Corrugated:** Wrinkled, folded.

**Cortex:** A maximal portion of ground tissue between the vascular system and the epidermis in a plant.

**Corymb:** A grouping of flowers where all the flowers are at the same level, the flower stalks of different lengths forming a flat-topped flower cluster, a broad, flat-topped inflorescence in which the flower stalks arise from different points on the main stem and the marginal flowers are the first to open.

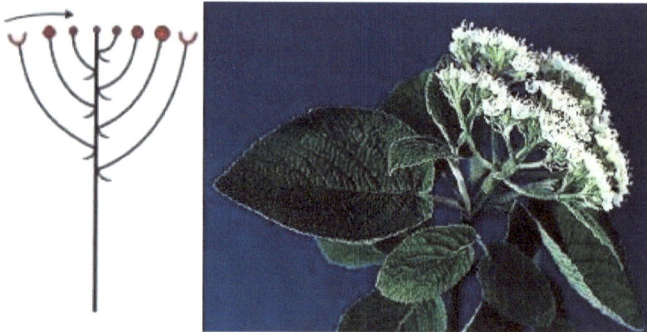

**Costa:** A portion of plant tissue that is a single or double strand in the center of a non-vascular leaf and has as part hydrome or leptome.

**Costate:** Ribbed, having longitudinal elevations.

**Cotyledon adaxial epidermis:** The adaxial/upper epidermal cell layer of the cotyledon.

**Cotyledon anlagen:** A phyllome anlagen that will give rise to a cotyledon primordium and is part of a peripheral zone of an embryo shoot apical meristem.

**Cotyledon epidermis:** A leaf epidermis that is part of a cotyledon.

**Cotyledon margin:** The margin of a cotyledon.

**Cotyledon primordium:** A phyllome primordium that develops from cotyledon anlagen and is part of a vegetative shoot apex and is committed to the development of a cotyledon.

**Cotyledon vascular system:** A vascular system that is part of a cotyledon.

**Cotyledon:** Seed leaves.

**Cotyledon:** A vascular leaf formed at the first shoot node of a plant embryo or a seedling.

**Cotyledonary node rhizoid:** An epidermal rhizoid that grows form a cotyledonary node.

**Cotyledonary node:** A stem node from which one or more cotyledons grow.

**Cover crop:** A crop which is planted in the absence of the normal crop to control weeds and add humus to the soil when it is plowed in prior to regular planting.

**Creeping:** Growing along the ground and producing roots at intervals along surface.

**Crenate:** With shallow roundish or bluntish teeth on the margin.

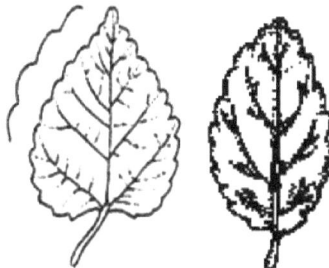

**Crenulate:** Similar to crenate but with smaller rounded teeth.

**Crisped:** Curled on the margin like a strip of bacon.

**Cristate:** With a terminal tuft or crest.

**Cross over:** The point along the meiotic chromosome where the exchange of genetic material takes place. This structure can often be identified through a microscope.

**Crossing over**: The reciprocal exchange of material between homologous chromosomes during meiosis, which is responsible for genetic recombination. The process involves the natural breaking of chromosomes, the exchange of chromosome pieces, and the reuniting of DNA molecules.

**Crown root:** Adventitious root formed at the base of the growing stem known as a crown.

**Crown:** The place where the roots and stem meet, which may or may not be clearly visible.

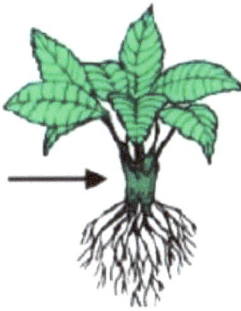

**Cruciform:** Cross-shaped.

**Crustaceous:** Dry and brittle.

**Cucullate:** Hooded or hood-shaped.

**Culm:** A hollow or pithy slender stem such as is found in the grasses and sedges.

**Cultivar:** A form of a plant derived from cultivation.

**Cultivate:** Process of breaking up the soil surface, removing weeds, and preparing for planting.

**Cultured plant callus:** A plant callus grown or maintained *in vitro*.

**Cultured plant cell:** A plant cell that is grown or maintained *in vitro*.

**Cultured plant embryo:** A plant embryo that is grown or maintained *in vitro*.

**Cultured somatic plant embryo:** A somatic plant embryo grown and maintained *in vitro*.

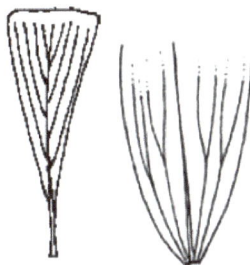

**Cultured zygote-derived plant embryo:** A zygotic plant embryo that is grown or maintained *in vitro*.

**Cuneate:** Triangular, stem attaches to point, Wedge-shaped, with the narrow part at the point of attachment.

**Cupule:** A cup-shaped involucre.

**Cuspidate:** Tipped with an abrupt short, sharp, firm point.

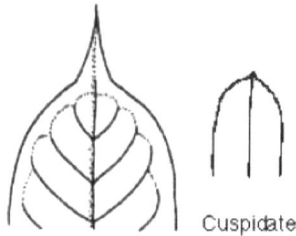

Cuspidate

**Cuticle:** A waxy membrane covering some leaves and roots that is water tight.

**Cuticular wax:** A portion of plant substance comprised predominantly of very long chain aliphatic lipids and is part of a plant cuticle.

**Cutin:** A portion of plant substance that is the inner layer of a cuticle composed of a polyester matrix of hydroxyl and hydroxy epoxy fatty acid C16 and C18 chains.

**Cuttings:** A method of propagation using sections of stems roots or leaves.

**Cyathiform:** Cup-shaped.

**Cyathium:** The specialized inflorescence characteristic of the *Euphorbiaceae*, consisting of a flower-like, cup-shaped involucre which carries the several true flowers within.

**Cyme:** A cluster of flowers were the end of each growing point produces a flower. New growth comes from side shoots and the oldest and first flowers to bloom are at the top, a broad, flat-topped inflorescence in which the central flower is the first to open.

# Glossary of Alphabet (D) Terminologies

**Damping off:** A fungus, usually affecting seedlings and causes the stem to rot off at soil level. It can also rot seeds before they even germinate. Sterilized seed starting mixes and careful sanitation practices can usually prevent this. Use care not to over-water. Provide good air circulation.

**Days to Harvest:** Days to harvest are usually indicated within the growing instructions for vegetables and fruit, and generally refer to the number of days it takes from setting out a transplant until the first harvest can be made.

**Days to Maturity:** Days to maturity are usually indicated within the growing instructions for vegetable seeds, and generally refer to the number of days it takes from sowing until the first harvest can be made. However, this is not a universally accepted definition, and may refer to the number of days it takes from setting out a transplant or seedling until the first harvest can be made.

**Dead heading:** Dead-heading is a simple type of pruning which involves the removal of a flower or cluster of flowers down to a leaf or set of leaves. The main reason for dead-heading is to discourage a plant from producing seeds. Often when a plant begins to set seeds, its natural tendency to produce flowers will diminish. By removing flowers before this takes place, one can disrupt the seed-forming process, there by causing the plant to produce more flowers. Dead-heading is most effectively done when flowers begin to fade or drop petals.

**Deca:** A prefix meaning ten.

**Deciduous:** Falling away after its function is completed; a plant which is deciduous will shed all of its foliage at the end of the growing season. Deciduous plants will then produce a new set of leaves at the onset of the next growing season.

**Decumbent:** Stems that lie on the ground but have the ends turning upward.

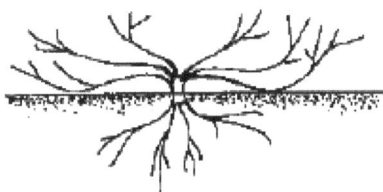

**Decurrent:** Adnate to the petiole or stem and extending downward, as a leaf base that extends downward along the stem.

**Decussate:** Arranged in pairs along the stem with each pair at right angles to the one above and below.

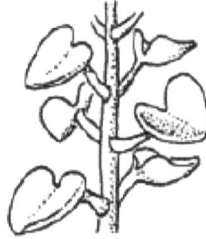

**Deflexed:** Bending downward or backward.

**Degenerate megaspore:** In monosporic and bisporic megasporogenesis: the megaspore that would not participate in mega gametogenesis.

**Dehiscence zone:** A portion of plant tissue that is part of a plant structure and consists of an arrow band of cells that undergoes dehiscence upon maturation of the structure.

**Dehiscent:** The way a fruit opens and releases its contents, normally in a regular and distinctive fashion, opening at maturity, opening spontaneously when ripe to discharge the seed content.

**Deltoid:** Broadly triangular in shape, stem attaches to side.

**Dense:** Congested, describing the disposition of flowers in an inflorescence.

**Dentate:** With sharp, outward-pointing teeth on the margin.

**Depauperate:** Starved or stunted, describing small plants or plant communities that are growing under unfavorable conditions.

**Derives by manipulation from:** A is a type of *in vitro* plant structure, and every instance of A exists at a point in time later than some instance of B from which it was created through human manipulation, and every instance of A inherited a biologically significant portion of its matter from the instance of B from which it was derived.

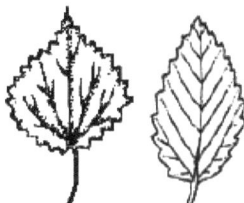

**Determinate growth:** Growing for a limited time, floral formation and leaves.

**Determinate nodule:** A root nodule characterized by a dividing infected cells and bacteria, vascular transfer cells absent and cell division is in the outer cortex.

**Determinate:** Describes an inflorescence in which the terminal flower blooms first, there by halting further elongation of the flowering stem.

**Dethatch:** Process of removing dead stems that build up beneath lawn grasses.

**Dextrorse:** Turned to the right or spirally arranged to the right.

**Di:** Prefix meaning two or twice.

**Diadelphous:** Having united filaments so that they are arranged in two groups.

**Diandrous:** Having two stamens.

**Diaphragmed pith:** Pith in which plates or nests of sclerenchyma may be interspersed with the parenchyma.

**Diarch protoxylem:** A portion of root exarch protoxylem tissue in which the primary xylem of the root system has two protoxylem strands or poles, and differentiation progresses centripetally, with the oldest elements farthest from the center of the plant axis.

**Dibble stick:** A pointed tool used to make holes in the soil for seeds, bulbs, or young plants.

**Dichogamy:** Flowers that cannot pollinate themselves because pollen is produced at a time when the stigmas are not receptive of pollen.

**Dichotomous:** branching regularly and repeatedly in pairs.

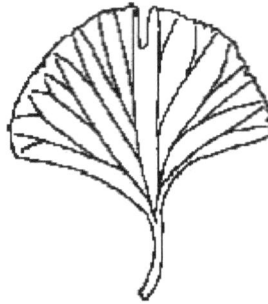

**Dicotyledon:** A plant having two seed leaves, one of the two major divisions of flowering plants.

**Didynamous:** Twinned, being in pairs, with two pairs of stamens of unequal length.

**Diffuse:** Loosely branching or spreading.

**Digitate:** Divided into finger-like lobes, Radiating from a common point, having a fingered shape, *i.e.* a shape like an open hand.

**Digynous:** Having two pistils.

**Dimorphic:** Having two forms.

**Dioecious plant:** Plants bear male flowers on one plant and female flowers on another. In order to produce fruit and viable seeds, both a female and male plant must be present.

**Disciform:** Having a flowering head that contains both filiform and disk flowers, referring to members of the *Asteraceae.*

**Discoid:** Having only disk flowers, referring to flower heads in the *Asteraceae.*

**Disjunct:** Separated from the main distribution of the population.

**Disk flower:** A small flower in the center of a head-type inflorescence.

**Disk:** An enlargement or out growth from the receptacle of the flower, located at the center of the flower of various plants. The term is also used for the central areas of the head in composites where the tubular flowers are attached.

**Dissected:** Finely cut or divided into many, narrow segments.

**Distal:** The end opposite the point of attachment, away from the axis.

**Distichous:** Two-ranked, that is with leaves on opposite sides of a stem and in the same plane.

**Distinct:** Having separate, like parts, those not at all joined to each other, often describing the petals on a flower.

**Disturbed:** Referring to habitats that have been impacted by the actions of people.

**Diurnal:** Growing in the daytime.

**Divaricate:** Widely diverging or spreading apart.

**Divergent:** Diverging or spreading.

**Divided:** Cut deeply, nearly or completely to the midrib.

**Dividing:** The process of splitting up plants, roots and all that have began to get bound together.This will make several plants from one plant, and usually should be done to mature perennials every 3 to 4 years.

**DNA**: An abbreviation for "deoxyribose nucleic acid", the carrier molecule of inherited genetic information.

**Dodeca:** Prefix meaning twelve.

**Domestication**: The process by which plants are genetically modified by selection over time by humans for traits that is more desirable or advantageous for humans.

**Dormancy:** The yearly cycle in a plants life when growth slows and the plant rests.Fertilizing should be withheld when a plant is in dormancy.

**Dormant:** A plant that is dormant is essentially resting. It is at a point in time when it cannot produce growth, usually due to climatic factors. Temperatures near freezing will cause the majority of landscape plants to enter a dormant state. It is at this time when most plants can be pruned, divided or transplanted most successfully.

**Dorsal:** Referring to the back or outer surface.

**Dorsifixed:** Attached at the back.

**Double digging:** Preparing the soil by systematically digging an area to the depth of two shovels.

**Double flower:** A flower with many overlapping petals which gives it a very full appearance.

**Drip line:** The circle which would exist if you draw a line below the tips of the outer most branches of a tree or plant.

**Drooping:** Erect or spreading at the base, then bending downwards.

**Drupe:** A fleshy indehiscent fruit enclosing a nut or hard stone containing generally a single seed such as a peach or cherry.

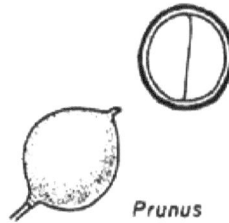

*Prunus*

**Dwarf:** Dwarf plants are ones which appear much smaller than members of the same species.This is often achieved by grafting a stem of a desirable variety to a root stock of a different variety. The difference in root stock and top allows the plant to grow only to a fraction of its usual height and width, allowing it to be grown in a smaller space than its full sized counterpart.

**Dwarfness:** The genetically controlled reduction in plant height. For many crops, dwarfness, as long as it is not too extreme, is an advantage, because it means that less of the crop's energy is used for growing the stem. Instead, this energy is used for seed or fruit or tuber production. The Green Revolution wheat and rice varieties were based on dwarfing genes.

# Glossary of Alphabet (E) Terminologies

**Early blossom:** Blooming plants that flourish in temperatures down to 35 degrees.

**Early wood:** The portion of the annual ring that is formed during the early phase of a tree's growth.

**Ebeneous:** Black.

**Ecad:** A plant assumed to be adapted to a specific habitat.

**Eccentric:** Off-center, not positioned directly on the central axis.

**Echinate:** Prickly.

**Ecotone:** The boundary that separates two plant communities, generally of major rank trees in woods and grasses for example in savanna.

**Ecotype:** Those individuals adapted to a specific environment or set of conditions.

**Ectogenesis:** Variation in plants due to conditions out side of the plants.

**Ectoparasites:** A parasitic plant that has most of its mass outside of the host, the body and reproductive organs of the plant live outside of the host.

**Elliptic:** Broadest near the middle and tapering gradually to both ends.

**Elongate:** Stretched out, many times longer than broad.

**Emarginate:** With a shallow notch at the apex.

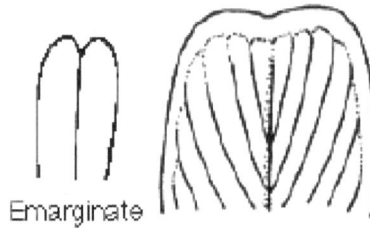

**Endemic:** Confined to a limited geographic area.

**Endocarp:** The inner layer of the pericarp, which is the wall of the ripened ovary or fruit.

**Ensiform:** Sword-shaped, as applied to a leaf.

**Entire:** Describing a leaf that has a continuous, unbroken margin with no teeth or lobes.

**Entomophilous:** Insect pollinated.

**Ephemeral:** Describes a plant or flower that lasts for only a short time or blooms only occasionally when conditions are right.

**Epicormic:** Vegetative buds that lie dormant beneath the bark, shooting after crown disturbance.

**Epidermis:** A layer of cells that cover all primary tissue separating them from the outside environment.

**Epigeal:** Living on the surface of the ground. See also terms for seeds.

**Epigean:** Occurring on the ground.

**Epigeic:** Plants with stolons on the surface of the ground.

**Epigeous:** On the ground. Used for leaf fungus that live on the surface of the leaf.

**Epigynous:** Flowers are present above the ovary, with stamens, pistils, and sepals attached tothe top of the ovary.

**Epilithic:** Growing on the surface of rocks.

**Epipetalous:** Born on the corolla, often used in reference to stamens attached to the corolla.

**Epiphloedal:** Growing on the bark of trees.

**Epiphloedic:** An organism that grows on the bark of trees.

**Epiphyllous:** Growing on the leaves.

**Epiphyte:** A plant which grows on another plant but gets its nourishment from the air and rain fall. They do no damage to the host plant.

**Epiphytic:** Having the nature of an epiphyte.

**Equinoctial:** A plants that has flowers that open and close at definite times during the day.

**Erect:** Growing up right.

**Erose:** Having an irregular margin as if it has been gnawed.

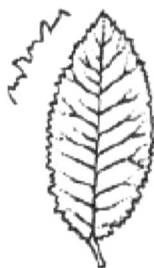

**Erosion:** The wearing away, washing away, or removal of soil by wind, water or man.

**Escape:** Plant originally under cultivation that has become wild, garden plant growing in natural areas.

**Espalier:** The process of training a tree or shrub so that its branches grow in a flat, exposed pattern by tying, pinching and pruning the branches.

**Eupotamous:** Living in rivers and streams.

**Euryhaline:** Normally living in salt water but tolerant of variable salinity rates.

**Eurythermous:** Tolerant of a wide range of temperature.

**Evanescent:** Fleeting, lasting for only a short time.

**Evaporation:** Process by which water returns to the air. Higher temperatures speed the process of evaporation.

**Even-pinnate:** A pinnately compound leaf ending in a pair of leaflets.

**Evergreen:** Evergreen plants retain their foliage throughout the year. For many ever greens, older interior foliage will begin to be shed with the onset of new growth. It is important to water ever greens thoroughly during winter.Especially during periods of dry and windy weather.

**Exclusive species:** Confined to specific location.

**Exfoliating:** Peeling off in thin layers or flakes.

**Exocarp:** The outer layer of the pericarp of a fruit.

**Exotic:** Not native to the area or region, introduced from another area.

**Exserted:** Sticking out past the corolla, the stamens protrude past the margin of the corolla lip.

**Exsiccatus:** A dried plant, most often used for specimens in a herbarium.

**Extant:** Still surviving, not completely extinct.

**Extirpated:**Destroyed or no longer surviving in the area being referred to, but may survive outside of that area.

**Extrorse:** Turned or opening outward away from the axis.

**Exudate:** A substance exuded or secreted from a plant.

**Eye:** An undeveloped bud growth which will ultimately produce new growth.

# Glossary of Alphabet (F) Terminologies

**Falcate:** Scimitar- or sickle-shaped.

**Farinose:** Covered with a mealy or whitish powdery substance.

**Fascicle:** A small cluster or bundle, a fairly common leaf arrangement.

Fasicled

**Fastigiate:** Clustered, parallel and erect, having a broom-like appearance.

**Feature:** Feature plants are visually dramatic plants which lend themselves perfectly for use as focal points in the landscape. They should be placed in a prominent area so that their special form or unique characteristic can be admired. Examples of feature plants include unusual weeping specimens, topiaries (sculptured plants), as well as trees with interesting bark or weeping branches.

**Fenestrate:** With small slits or areas thinned so as to be translucent.

**Ferruginous:** Rust-colored.

**Fertile:** Having the capacity to produce fruit, having a pistil.

**Fertilizer:** Organic or inorganic plant foods which may be either liquid or granular used to amend the soil in order to improve the quality or quantity of plant growth.

**Fibrous:** Describes roots that are thread-like and normally tough.

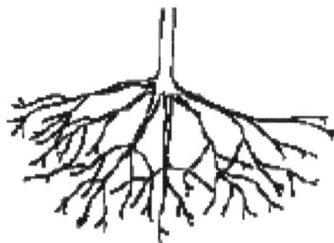

**Filament:** The basal, sterile portion of a stamen below the anthers.

**Hasnain Nangyal**

**Filiform:** Threadlike; A type of flower in the *Asteraceae* which is pistillate and has a very slender, tubular corolla.

**Fimbriate:** Finely cut into fringes, the edge of a frilly petal or leaf.

**Fistulose:** Hollow like a tube or pipe.

**Flabellate:** Fan-shaped, as in a fan-shaped structure.

**Flaccid:** Soft and weak, limp.

**Flange:** A projecting rim or edge.

**Flat:** A shallow box or tray used to start cuttings or seedlings.

**Flavescent:** Yellowish.

**Fleshy:** Describes roots that are relatively thick and soft, normally made up of storage tissue. Roots are typically long and thick but not thickly rounded in shape.

**Flexuose or flexuous:** With curves or bends, somewhat zig zagged.

**Floating row cover:** A light weight fabric that is spread or floated over a row of plants to trap heat during the day and release it at night. Use them to get a jump start in the spring, fend off pests, and extend your fall growing season.

**Floc:** A tuft of soft, woolly hair.

**Floccose:** Wooly, covered with soft wooly tufted hairs that are usually easily rubbed off.

**Floret:** A small individual flower in a flower head.

**Floricane:** The second-year flowering and fruiting cane or shoot of *Rubus*.

**Flower bud:** A bud from which only a flower or flowers develop.

**Flowering:** Flowering landscape plants produce beautiful displays of colorful blooms. Use a single flowering specimen to brighten a shady spot or plant in mass for a burst of color. Utilize Varieties with different bloom times to brighten the landscape all season long. Plant flowering plants in highly visible areas where they can be enjoyed. Utilize green-leafed plants as a back ground to make the flowers stand out where possible.

**Fluted:** With furrows or grooves.

**Foliar feeding:** Fertilizer applied in liquid form to the plants foliage in a fine spray.

**Foliolate:** Having leaflets.

**Follicle:** A dry, many-seeded fruit derived composed of a single carpel l and opening along one side only like a milk weed pod.

**Forb:** A non-grass like herbaceous plant.

**Forcing:** The process of hastening a plants growth to maturity or bloom.

**Foundation or hedge:** Foundation or hedge plants are perfect for softening the harsh edges associated with buildings. They are also useful to assort favorable elements of the home and hide less desirable ones. Many foundation plants provide a natural backdrop to shorter foreground plantings. Hedges are used to create lush green barriers between yards and to increase privacy. Foundation plantings help to reduce the amount of energy used in a home by insulating the house from wind and sun.

**Fovea:** A small pit or depression.

**Frond:** The term used to describe the branch and leaf structure of a fern or members of the palm family.

**Frost:** The condensation and freezing of moisture in the air. Tender plants will suffer extensive damage or die when exposed to frost.

**Fructiferous:** Fruit-bearing.

**Fruit:** A structure contains all the seeds produced by a single flower.

**Frutescent:** Shrubby or bushy in the sense of being woody.

**Fruticose:** Woody stemmed with a shrub-like habit. Branching near the soil with woody based stems.

**Fugacious:** Lasting for a short time: soon falling away from the parent plant.

**Fulvous:** Dull yellowish-brown or yellowish-gray, tawny.

**Funicle:** The stalk that connects the ovule to the placenta.

**Funnel form:** Gradually widening upwards, as in the flowers of morning glory.

**Furcate:** Forked, dividing into two divergent branches.

**Furfuraceous:** Scurfy, branlike, and flaky.

**Fuscous:** Dark grayish-brown, dusky.

**Fusiform:** Spindle-shaped, thickest in the middle and drawn out at both ends.

# Glossary of Alphabet (G) Terminologies

**Gall:** Outgrowth on the surface caused by invasion by other lifeforms, such as parasites.

**Geniculate:** Bent abruptly like a knee or a stove pipe.

**Germinate:** The process that transforms the embryo within a seed into a seedling.

**Girdling:** The choking of a branch by a wire or other material, most often in the stems of woody plants that have been tied tightly to a stake or support.

**Glabrate:** Becoming glabrous in age.

**Glabrous:** Smooth without any pubescence at all.

**Gland:** A depression or pro tuberance that exists for the purpose of secreting.

**Glandular:** Producing tiny globules of sticky or oily substance.

**Glandular-punctate:** Covered across the surface with glands.

**Glans:** A dry dehiscent fruit borne in a cupule.

**Glaucescent:** Slightly glaucous.

**Glaucous:** Covered with a thin, light-colored waxy or powdery bloom.

**Globose:** Globe-shaped, spherical.

**Glochids:** Barbed bristles on cacti.

**Glomerate:** Crowded, congested or compactly clustered.

**Glume:** In grasses, the bracts that form the lowermost parts of the spikelet.

**Glutinous:** Having a sticky surface.

**Gonioautoicous:** Male is bud-like in the axil of a female branch.

**Gracile:** Slender and graceful.

**Graft:** To graft is to join together the strong growing roots of one variety of plant with another desired variety's stem for top growth. The resulting union allows for a more vigorous plant than the desired variety would have had if grown on its own roots. When planting grafted plants it is important to remember that the point of the graft should always be planted higher than the existing grade. Any growth which might occur below the graft is usually undesirable and should be removed immediately.

**Grafting:** The uniting of a short length of stem of one plant onto the root stock of a different plant. This is often done to produce a hardier or more disease resistant plant.

**Grain:** The fruit of grasses.

**Gregarious:** Growing in groups or colonies.

**Grenadine:** Bright red.

**Ground cover:** A group of plants usually used to cover bare earth and create a uniform appearance.

**Growing season:** The number of days between the average date of the last killing frost in spring and the first killing frost in fall. Vegetables and certain plants require a minimum number of days to reach maturity, so be sure your growing season is long enough.

**Guarantee:** If any tree, shrub or perennial you buy at Lowe's doesn't survive a year, we will replace it. Just bring it in with the receipt.

**Guard cell:** One of the paired epidermal cells that control the opening and closing of a stoma in plant tissue.

**Gynandrium:** Combined male and female structure.

**Gynobase:** An elongation or enlargement of the receptacle that supports the carpels or nutlets, as in many species of the *Boraginaceae*.

**Gynodioecy:** Describes a plant species or population that has some plants that are female and some plants that are hermaphrodites.

**Gynoecium:** The whorls of carpels. May comprise one or more pistils. Each pistil consists of an ovary, style and stigma.

# Glossary of Alphabet (H) Terminologies

**Habit:** The overall appearance of a plant.

**Halophyte:** A plant that can tolerate an abnormal amount of salt in the soil.

**Hamate:** Hook-shaped, hooked at the tip.

**Harden off:** The process of gradually acclimatizing green house or indoor grown plants in stages to different temperatures or to outdoor growing conditions.

**Hardiness:** The ability of a plant to withstand low temperatures or frost, without artificial protection.

**Hardpan:** The impervious layer of soil or clay lying beneath the top soil.

**Hastate:** Spear or arrowhead-shaped with the basal lobes facing outward.

**Haustorial:** Specialized roots that invade other plants and absorb nutrients from those plants.

**Heading back:** Cutting an older branch or stem back to a stub or twig.

**Heartwood:** The older, nonliving central wood of a tree or woody plant, usually darker and harder than the younger sapwood also called duramen.

**Heeling in:** Temporarily setting a plant into a shallow trench and covering the roots with soil to provide protection until it is ready to be permanently planted.

**Helicoid:** Coiled spirally like a spring or a snail shell.

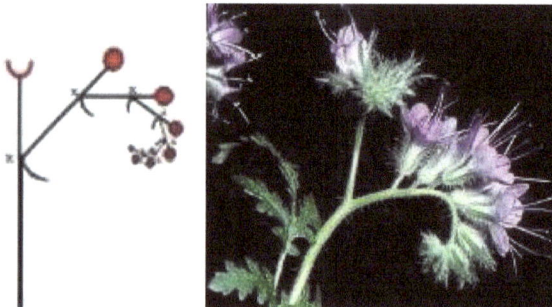

**Heliotropic:** The movement of plant parts in response to a light source.

**Hemiparasite:** A plant that derives its energy both from parasitism and from photosynthesis.

**Herbaceous:** Non-woody and dying to the ground at the end of the growing season. Annual plants die, while perennials regrow from parts on the soil surface or below ground the next growing season.

**Heteromorphic:** Of one or more kind or form.

**Heterostylous:** Having different kinds of style lengths.

**Hexa:** A prefix meaning six.

**Hibernal:** Flowering or appearing in the winter.

**Hilling:** Hilling is a method of protecting roses from harsh winter weather. Before the onset of cold weather, pile a soil and compost mix about one foot high, making sure to cover the point where the plant was grafted. Fallen tree leaves or any suitable mulch should be added on top of the soil mix. Excessively tall canes should be pruned if in a windy location to prevent movement that could cause the hill to become unstable.

**Hilum:** A scar on a seed indicating its point of attachment.

**Hip:** A fleshy, berry-like fruit, as in some members of the *Rosaceae*.

**Hirsute:** Pubescent with long shaggy hairs, often stiff or bristly to the touch.

**Hirtellous:** Pubescent with very small, coarse, stiff hairs.

**Hispid:** Rough-haired.

**Hoary:** Covered with white or gray, short, fine hairs.

**Holosericeous:** Covered with fine, silky hairs.

**Homogamous:** When the flower anthers and the stigma are ripe at the same time.

**Homomorphic:** All of the same kind or form.

**Honeydew:** The sticky secretion produced by sucking insects such as aphids.

**Hooked:** Abruptly curved at the tip.

**Host:** A plant providing nourishment to a parasite.

**Humifuse:** Spreading along or over the ground.

**Humistrate:** Lying on the ground.

**Humus:** The brown or black organic part of the soil resulting from the partial decay of leaves and other matter.

**Hyaline:** Thin, translucent or transparent.

**Hybrid:** The offspring of two plants of different species or varieties of plants. Hybrids are created when the pollen from one kind of plant is used to pollinate and entirely different variety, resulting in a new plant altogether.

**Hydrophilous:** Water pollinated, pollen is moved in water from one flower to the next.

**Hydrophytic:** Adapted to growing in water.

**Hydroponics:** The science of growing plants in mineral solutions or liquid, instead of in soil.

**Hypanthium:** A cup-shaped enlargement of the receptacle, creation by the fusion of sepals, petals and stamens.

**Hypogynous:** Flowers are present below the ovary.

# Glossary of Alphabet (I) Terminologies

**Imbricate:** Overlapping, like shingles on a roof.

**Imperfect:** Describes a flower that has stamens or pistils but not both.

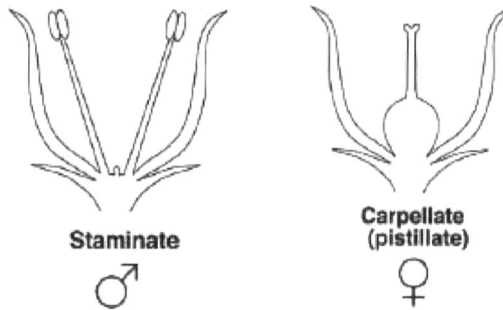

**Staminate**
♂

**Carpellate
(pistillate)**
♀

**Implicate:** Twisted together, intertwined.

**Incised:** Cut, often deeply, usually irregularly, but seldom as much as one-half the distance to the midrib or base.

**Included:** Not exerted or protruding beyond the surrounding organ.

**Incobus:** Describing the arrangement of leaves of a liverwort, contrast with succubous.

**Incumbent:** A term referring to seeds in which the embryonic root is wrapped around and lies adjacent to the back of one of the two cotyledons.

**Indehiscent:** Fruits that do not have specialized structures for opening and releasing the seeds, they remain closed after the seeds ripen and are opened by animals, weathering, fire or other external means.

**Indeterminate growth:** Inflorescence and leaves growing for an indeterminate time, until stopped by other factors such as frost.

**Indeterminate:** Describes an inflorescence in which the outer or lower flowers bloom first, allowing an indefinite elongation of the flowering stem.

**Indigenous:** Native to an area.

**Indurate:** Hardened or stiffened.

**Indusium:** A scale-like outgrowth on a fern leaf which forms a covering for the sporangia.

**Inferior ovary:** One that is situated below the point of attachment of the sepals and petals, and possibly below the point of attachment of all other flower parts and embedded in the floral stem.

**Inflexed:** Turned abruptly or bent inwards.

**Inflorescence:** Cluster of flower, the flowering portion of a plant.

**Infra:** A prefix meaning below or beneath.

**Inframedial:** Below the middle.

**Infraspecific:** Below the species level.

**Infundibular:** Funnel-shaped.

**Innate:** Borne at the apex.

**Inter:** A prefix meaning between or among.

**Internal shoots:** Internal shoots are vertical branches within the structure of a plant.

**Internode:** Spaces between the nodes, the portion of a stem between two successive nodes.

**Interrupted:** Not continuous, with gaps.

**Introrse:** Turned or opening inward toward the axis as an anther toward the center of a flower.

**Involucel:** A secondary involucre as in the *Apiaceae.*

**Involucre:** A tube of thallus tissue that protects the archegonia, a set of bracts subtending a flower or an inflorescence.

**Involute:** With both edges inrolled toward the mid nerve on the upper surface.

**Irregular:** Describes a flower that is not radially symmetric, the similar parts of which are unequal in size or form.

# Glossary of Alphabet (J) Terminologies

**Joint:** The point on a plant stem from which a leaf or leaf-bud grows, more commonly termed anode.

**Jugate:** With parts in pairs.

**Junciform:** Rush-like in appearance.

# Glossary of alphabet (K) Terminologies

**Keel:** The two lower petals of most pea flowers united or partially joined to form a structure similar to the keel of a boat.

# Glossary of Alphabet (L) Terminologies

**Labiate:** Lipped.

**Lacerate:** Irregularly cut or cleft.

**Laciniate:** Cut into slender lobes.

**Lacunate:** Pitted.

**Lacustrine:** Growing around lakes.

**Laevigate:** Lustrous, shining.

**Lanate:** With woolly hairs. Thick wool like hairs.

**Lanceolate:** Significantly longer than wide and widest below the middle, gradually tapering toward the apex.

**Lanulose:** With very short hairs, minutely downy or wooly.

**Late wood:** The portion of the annual ring that is formed after formation of early wood has ceased.

**Latent bud:** An axillary bud whose development is inhibited, sometimes for many years, due to the influence of apical and other buds Also known as dormant bud.

**Lateral buds:** A bud located on the side of the stem, usually in a leaf axil.

**Lath:** In gardening, an overhead structure of evenly spaced slats of wood or other materials used to create shade.

**Latifoliate:** With broad leaves.

**Lax:** Non upright, growth not strictly upright or hangs down from the point of origin.

**Layering:** A method of propagation, by which a branch of a plant is rooted while still attached to the plant by securing it to the soil with a piece of wire or other means.

**Leaching:** The removal or loss of excess salts or nutrients from soil. The soil around over fertilized plants can be leached clean by large quantities of fresh water used to wash the soil. Areas of extremely high rainfall sometimes lose the nutrients from the soil by natural leaching.

**Leaf axils:** The space created between a leaf and its branch. This is especially pronounced on monocots like bromeliads.

**Leaf bud:** A bud that produces a leafy shoot.

**Leaf mold:** Partially decomposed leaf matter, used as a soil amendment.

**Leaf scar:** The mark left on a branch from the previous location of a bud or leaf.

**Leaf:** The photosynthetic organ of a plant that is attached to a stem, generally at specific intervals.

**Leaflet:** A separate blade among others comprising a compound leaf.

**Legume:** A dry, dehiscent fruit derived from a single carpel and usually opening along two lines of dehiscence like a pea pod.

**Lemma:** In grasses, the lower and usually larger of the two bracts of the floret.

**Lenticels:** One of the small, corky pores or narrow lines on the surface of the stems of woody plants that allow the inter change of gases between the interior tissue and the surrounding air.

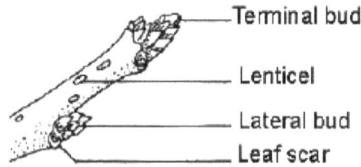

**Lepidote:** Covered with small scurfy scales.

**Liana:** An herbaceous or woody usually perennial, climbing vine that roots in the ground and is characteristic especially of tropical forests.

**Ligneous:** woody.

**Lignotuber:** Root tissue that allows plants to regenerate after fire or other damage.

**Ligulate:** Describing a floral head in the *Asteraceae* that contains only ray flowers, or ligules, strap-shaped.

**Limb:** The upper, expanded portion of a corolla which has fused petals.

**Linear:** Long and narrow with sides that are parallel or nearly.

**Lineate:** Marked with parallel lines.

**Lingulate:** Tongue-shaped.

**Littoral:** Growing along the shore.

**Livid:** Pale grayish-blue.

**Loam:** A rich soil composed of clay, sand, and organic matter.

**Lobe:** Usually a rounded segment of an organ.

**Lobulate:** With small lobes.

**Locule:** A cavity of the ovary which contains the ovules.

**Loculicidal:** Said of a capsule, longitudinally dehiscent through the ovary wall at or near the center of each chamber or locule.

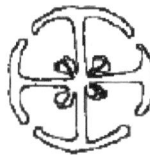

**Loculus:** The cavities located within a carpel, ovary or anther.

**Loment:** A legume which is constricted between the seeds.

**Lunate:** Crescent-shaped.

**Lurid:** Pale brown to yellowish-brown.

**Lutescent:** Yellowish.

**Lyrate:** Lyre-shaped, pinnatifid with the terminal segment large and rounded and the lower lobes increasingly smaller toward the base.

# Glossary of Alphabet (M) Terminologies

**Machaerantheroid:** Having involucral bracts with recurved tips.

**Macro:** prefix meaning large or long.

**Macrophyllous:** Having large leaves, Leaves with a branching vascular system.

**Maculate:** Spotted or blotched.

**Malacophilous:** Pollinated by snails and slugs.

**Malacophyllous:** With soft leaves.

**Mammilate:** With nipple-like pro tuberances.

**Manicate:** With a thick, interwoven pubescence.

**Manure:** Organic matter, excreted by animals, which is used as a soil amendment and fertilizer.Green manures are plant cover crops which are tilled into the soil.

**Marcescent:** Withering but still persistent as with petals and sepals or the basal leaves of some plants.

**Margin:** The edge, as of a leaf blade.

**Marginate:** Distinctly margined.

**Marginal placentation:** Ovules are attached to the folded margins of the carpel, giving the appearance that there is only one elongated placenta on one side of the ovary. Can only is found in a simple pistil. This is conspicuous in legumes.

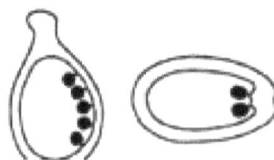

**Matinal:** Blooming in the early morning.

**Mauve:** Bluish or pinkish-purple.

**Mealy:** Describing a surface that is covered with minute, usually rounded particles.

**Mega:** Prefix meaning large.

**Membranous:** Thin, flexible and more or less translucent, like a membrane.

**Merous:** A suffix utilized to indicate the number of parts or divisions in a particular structure or organ, as in 4-merous or 4-parted.

**Mesic:** Describes a habitat that is generally moist throughout the growing season.

**Meso:** Prefix meaning middle.

**Mesocarp:** The middle layer of the pericarp of a fruit.

**Mesophytic:** Adapted to growing under medium or average conditions, especially relating to water supply.

**Micro nutrients:** Mineral elements which are needed by some plants in very small quantities. If the plants you are growing require specific 'trace elements' and they are not available in the soil, they must be added.

**Micro:** Prefix meaning small.

**Microclimate:** Variations of the climate within a given area, usually influenced by hills, hollows, structures or proximity to bodies of water.

**Microphyllous:** Bearing small leaves.

**Midrib:** The central vein of the leaf blade.

**Midvein:** The central vein of a leaflet.

**Mixed:** Buds that have both embryonic flowers and leaves.

**Monadelphous:** Having stamens with filaments united in a single group, bundle or tube.

**Monandrous:** With a single stamen.

**Monanthous:** One flowered.

**Mono:** Prefix meaning one

**Monocarpic:** Plants that live for a number of years then after flowering and seed set die.

**Monocotyledon:** A plant having only one seed leaf.

**Monodelphous:** Stamen filaments united into a tube.

**Monoecious:** A monoecious plant produces both female and male flowers on a single plant. The plant will be able to produce berries or fruit without the need of another plant. However, in some instances berry or fruit production may be increased by planting multiples of the same variety.

**Monotypic:** Describing a genus that contains only a single species.

**Montane:** Of or pertaining to, or growing in, the mountains.

**Mounding**: Mounding plants grow in such a way as to produce growth both vertically and horizontally, creating a rather rounded appearance. Mounding plants can serve as a transition in the landscape between strongly upright and low, trailing plants.

**Mucilaginous:** Slimy and moist.

**Mucronate:** Having a short projection at the tip, as of a leaf.

**Mulch:** Any loose material placed over the soil to control weeds and conserve soil moisture. Usually this is a coarse organic matter, such as leaves, clippings or bark, but plastic sheeting and other commercial products can also be used.

**Multi:** Prefix meaning many.

**Multifid:** Cleft into very many narrow lobes or segments.

**Multiflorus:** Many flowered.

**Multifoliate:** Bearing many leaves.

**Muricate:** Rounded or roughened with short, hard or warty points.

**Mycorrhizal:** Having a symbiotic relationship between a fungus and the root of a plant.

# Glossary of Alphabet (N) Terminologies

**Nacreous:** Having a pearly luster.

**Napiform:** Turnip-shaped.

**Nascent:** In the process of being formed.

**Natant:** Floating in water.

**Native plant:** Any plant that occurs and grows naturally in a specific region or locality.

**Naturalize:** To plant randomly, without a pattern. The idea is to create the effect that the plants grew in that space without man's help, such as you would find wild flowers growing.

**Navicular:** Boat-shaped.

**Nectar disk:** When the floral disk contains nectar secreting glands, often modified as its main function in some flowers.

**Nectar:** A fluid produce by nectaries high in sugar content, used to attract pollinators.

**Nectary:** A gland that secrets nectar, most often found in flowers but also produced on other parts of plants too.

**Netted:** Same as reticulated, in the form or pattern of a network.

**Neuter:** Lacking a pistil or stamens.

**Nidulent:** Lying within a cavity, embedded within a pulp.

**Nigrescent:** Blackish.

**Nitid:** Lustrous, shining.

**Niveous:** White.

**Nodding:** Hanging down.

**Node:** Where leaves and buds are attached to the stem, A point on a stem where leaves or branches originate.

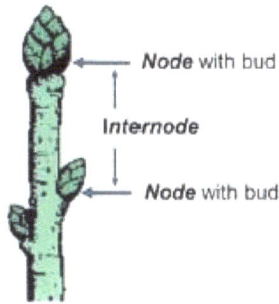

**Nodose:** Knobby or knotty.

**Nomophilous**: Growing in or loving pastures.

**Notate:** Marked with lines or spots.

**Numerous:** Eleven or more, same as many.

**Nut:** A dry, usually one-seeded, indehiscent fruit with a hard-walled exterior.

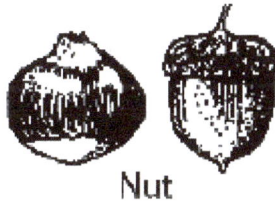

Nut

**Nutant:** Nodding, drooping.

**Nutlet:** A small nut or one of the sections of the mature ovary of some members of the *Verbenaceae* or *Lamiaceae.*

**Nyctagimous:** Opening at night.

**Nyctanthous:** Night flowering.

# Glossary of Alphabet (O) Terminologies

**Obconic:** Cone shaped and attached at the pointed end.

**Obcordate:** Heart-shaped, stem attaches to tapering point.

**Oblanceolate:** Top wider than bottom.

**Obligate:** Restricted to particular conditions or circumstances.

**Oblique:** With side's unequal, usually describing the base of a leaf.

**Oblong:** Two to four times longer than broad with nearly parallel sides, but broader than linear.

**Obovate:** Teardrop shaped, stem attaches to tapering point.

**Obtuse:** Blunt or rounded at the apex.

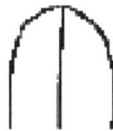

**Obverse:** Describing a leaf that is narrower at the base than at the apex.

**Obvolute:** A vernation in which two leaves are overlapping in the bud in such a manner that one-half of each is external and the other half is internal, *i.e.* each leaf both overlaps the next and is in turn overlapped by the one before.

**Ochreoleucous:** Yellowish white, cream-colored.

**Ocrea:** A sheath around the stem derived from the leaf stipules, primarily used inthe *Polygonaceae.*

**Octo:** Prefix meaning eight.

**Odd pinnate:** Describing a pinnately-compound leaf with a single terminal leaflet.

**Oligomeris:** With less than the typical number of parts.

**Oligophyllous:** With few leaves.

**Olivaceous:** Olive-green.

**Opposite:** Leaves are arranged in pairs on opposite sides of the branch.

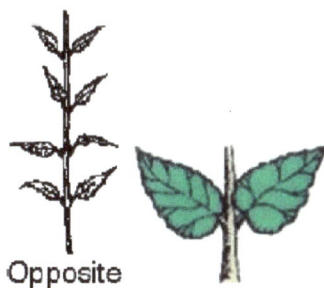

Opposite

**Orbicular:** Circular.

**Organic gardening:** The method of gardening utilizing only materials derived from living things.

**Organic material:** Any material which originated as a living organism. *i.e.* peat moss, compost, manure.

**Ornithophillous:** Bird-pollinated.

**Orophilous**: Growing in or preferring mountain areas.

**Orthotropic growth:** Growth in a vertical direction.

**Oval:** Broadly elliptic, the width over half the length.

**Ovary:** The basal portion of a pistil where female germ cells develop into seeds after germination.

**Ovate:** Egg shaped, wider below the middle.

**Ovoid:** An egg-shaped solid.

**Ovule:** The structure that develops into the seed inside the ovary.

# Glossary of Alphabet (P) Terminologies

**Pachyphyllous:** With thick leaves.

**Palate:** An appendage or raised area on the lower lip of the corolla which partially blocks the throat.

**Palea:** In grasses, the upper and generally smaller of the two bracts of the floret.

**Pallid:** Pale.

**Palmate:** Radiating from a single point like the spreading fingers of an out stretched hand.

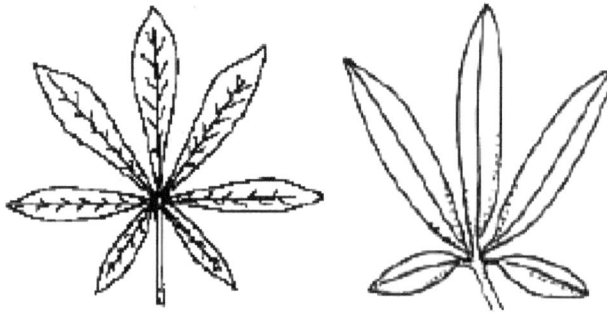

**Palmatifid:** Palmately cleft or lobed.

**Paludose:** Growing in wet meadows or marshes.

**Palustrine:** Same as paludose.

**Pandurate:** Fiddle-shaped.

**Panicle:** A raceme with branches and each branch having a smaller raceme of flowers. The terminal bud of each branch continues to grow, producing more side shoots and flowers.

**Pannose:** With a covering of short, dense, felty or wooly tomentum.

**Papilionaceous:** Describing the structure of a corolla, wings and keel.

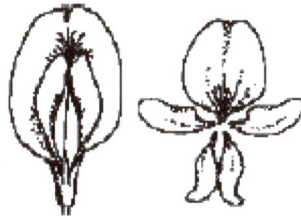

**Pappose:** Pappus bearing.

**Pappus:** Collectively, the bristles, hairs or scales at the apex of an achene in the *Asteraceae.*

**Paraphyses:** Sterile hairs surrounding the archegonia and antheridia.

**Parasite:** Organism which derives most or all of its food from another organisim to which it attaches itself.

**Parasitic:** A plant which lives on, and acquires it's nutrients from another plant.This often results in declined vigor or death of the host plant.

**Parietal placentation:** Ovules are attached to the side walls of the ovary such that an ovary usually has one locule and therefore no septa. Can only be found in a syncarpous gynoeium.

**Paripinnate:** Pinnate lacking a terminal leaflet.

**Parted:** Lobed or cut in over half-way and often very close to the base or midrib.

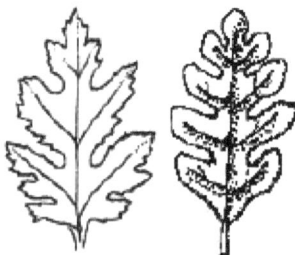

**Peat moss:** The partially decomposed remains of various mosses. This is a good, water retentive addition to the soil, but tends to add the acidity of the soil pH.

**Pectinate:** Describing a pinnatifid leaf whose segments are narrow and arranged like the teeth of a comb.

**Pedate:** Palmate, with cleft lobes.

**Pedicel:** The stem or stalk that holds a single flower in an inflorescence.

**Peduncle:** The part of a stem that bears the entire inflorescence, normally having no leaves orthe leaves are reducing to bracts. When the flower is solitary, it is the stem or stalk holding the flower.

**Pedunculate:** Having a peduncle.

**Peltate:** A type of leaf having its petiole attached to the center of the lower surface of the blade.

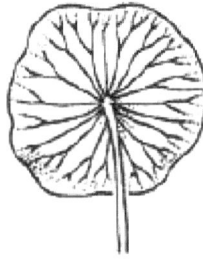

**Pendent:** Hanging downward or drooping.

**Penicillate:** With a tuft of short hairs at the end, like a brush.

**Penta:** Prefix meaning five.

**Pepo**: A fleshy, indehiscent fruit with a hard, more or less thickened rind and a single many-seeded locule, characteristic of the *Cucurbitaceae.*

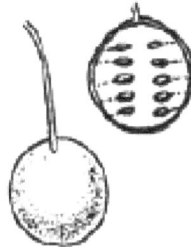

**Perennial:** A plant living for more than two years, a non-woody plant which grows and lives for more than two years. Perennials usually produce one flower crop each year, lasting anywhere from a week to a month or longer.

**Perfect:** Containing both stamens and pistils.

**Perfoliate:** The stem apparently piercing the leaf or surrounded by basally joined opposite leaves.

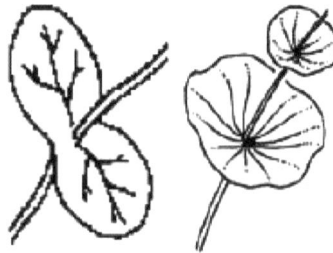

**Perianth:** A collective term for the calyx and corolla.

**Pericarp:** The outer wall of mature fruit, the body of the fruit from its outside surface to the chamber was the seeds are, including the outside skin of the fruit and the inside lining of the seed chamber.

**Perichaetium:** The cluster of leaves with the enclosed female sex organs.

**Perigonium:** The cluster of leaves with the enclosed male sex organs.

**Perigynous:** Situated around but not attached to the ovary directly, describing a flower whose stamens and pistils are joined to the calyx tube and the ovary is superior.

**Perlite:** A mineral, which when expanded by a heating process forms light granuals. Perlite is a good addition to container potting mixes, to promote moisture retention while allowing good drainage.

**Persistent:** Remaining attached after the usual time of falling.

**Pest:** Any insect or animal which is detrimental to the health and well-being of plants or other animals.

**Petal:** A single segment of a divided corolla.

**Petaloid:** Having the appearance of a petal.

**Petiole:** A leaf stalk supporting a blade and attaching to a stem at a node.

**Petiolule:** The leaf stalk of a leaflet.

**pH**: Basically, pH is a measure of the amount of lime contained in your soil. A soil with a pHlower than 7.0 is an acid soil, a soil pH higher than 7.0 is alkaline soil. Soil pH can be tested with an inexpensive test kit.

**Phloem:** The food conducting tissue of vascular plants, bark.

**Photosynthesis:** The internal process by which a plant turns sunlight into growing energy.The formation of carbohydrates in plants from water and carbon dioxide, by the action of sunlight on the Chlorophyll within the leaves.

**Phreatophyte**: A perennial plant that has deep and extensive root systems that enable it to tap underground sources of water.

**Phyllary:** One of the bracts below the flowerhead in the *Asteraceae.*

**Pilose:** Having long, soft, straight hairs.

**Pinching back:** Utilizing the thumb and forefinger to nip back the very tip of a branch or stem.Pinching promotes branching, and a bushier, fuller plant.

**Pinnate:** Two rows of leaflets, a compound leaf structure with a feather-like formation of leaflets arranged in pairs or alternating along the main stem.

**Pinnatifid:** So deeply cleft or cut as to appear pinnate.

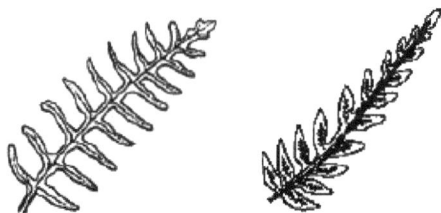

**Pinnatisect:** Cut, but not to the midrib.

**Pisaceous:** Pea green.

**Pistil:** The central reproductive organ of a flower, consisting of ovary, style and stigma.

**Pistillate:** A female flower that has two or more pistils but no functional stamens.

**Pith:** The spongy tissue at the center of a stem.

**Plagiotropic growth:** Growth inclined away from the vertical, inclined towards the horizontal.

**Planoconvex:** Flat on one side and rounded on the other.

**Plant Trunk:** The trunk is the woody, thickened main stem of a plant.

**Plumbeous:** Lead colored.

**Plumose:** Appearing plume like or feathery from fine hairs that line two sides of a central axis.

**Plumule:** The part of an embryo that give rise to the shoot system of a plant.

**Pod:** A dry dehiscent fruit containing many seeds. *i.e.* follicles and legumes.

**Pollination:** The transfer of pollen from the stamen to the pistil, which results in the formationof a seed. Hybrids are created when the pollen from one kind of plant is used to pollinate an entirely different variety, resulting in a new plant altogether.

**Pollinator:** A pollinator is the agent by which pollen from a flower is transported to accomplish the pollination process. Most plants grown for their edible produce require pollination to set fruit. Bees are common pollinators. Planting a wide variety of blooming plants can help attract natural pollinators in your area to your landscape.

**Poly:** Prefix meaning many.

**Polyandrous:** With many stamens.

**Polyanthous:** With many flowers.

**Polycephalous:** With many flower heads.

**Polygamous:** Having both unisexual and bisexual flowers on the same plant.

**Pome:** A fleshy indehiscent fruit derived from an inferior, compound ovary and consisting of a modified floral tube surrounding a core with several seeds, such as an apple.

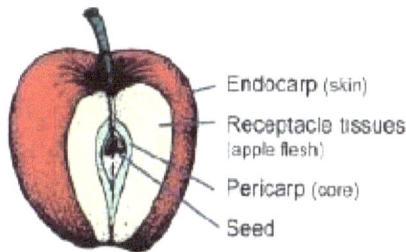

**Poricidal:** Anthers opening by terminal pores, like a poppy capsule.

**Posterior:** On the side next to the axis.

**Potting soil:** A soil mixture designed for use in container gardens and potted plants. Potting mixes should be loose, light and sterile.

**Precocious:** Flowering before the leaves emerge.

**Prickle:** An extension of the cortex and epidermis that ends with a sharp point.

**Primary:** Roots that develop from the radicle of the embryo, normally the first root to emerge from the seed as it germinates.

**Primocane:** The first-year cane or shoots of *Rubus*.

**Procumbent:** Growing prostrate or trailing but not rooting at the nodes.

**Progynous:** When the carpels mature before the stamens produce pollen.

**Propagation:** Various methods of starting new plants ranging from starting seeds to identical clones created by cuttings or layering.

**Prostrate:** Growing flat on the soil surface.

**Prostrate:** Laying flat on the ground, stems or even flowers in some species.

**Protandrous:** When pollen is produced and shed before the carpels are mature.

**Protogynous:** Describing a plant in which stigma receptivity precedes and does not overlap the period of pollen release.

**Proximal:** Nearest the axis or base.

**Pruning:** The cutting and trimming of plants to remove dead or injured wood or to control and direct the new growth of a plant.

**Pseudautoicous:** Dwarf male plants growing on living leaves of female plants.

**Pseudoperianth:** An involucre that resembles a perianth, but is made of thallus tissue, and usually forms after the sporophyte develops.

**Ptero:** Prefix meaning winged.

**Pterocarpous:** Winged fruits.

**Pterospermous:** With winged seeds.

**Ptyxis:** The way in which an individual leaf is folded within an unopened bud.

**Puberulent:** Minutely pubescent.

**Pubescent:** Covered with short, soft hairs.

**Pulverulent:** Dusty or chalky, as applied to the powdery coating on the stems and leaves of some plants.

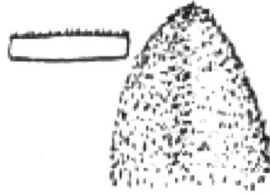

**Pulvinate:** Cushion or mat like.

**Pulvinus:** The swollen base of a petiole usually involved in leaf movements and leaf orientation.

**Punctate:** Dotted with pits or with translucent, sunken glands, or with colored dots.

**Pyriform:** Pear shaped.

# Glossary of Alphabet (Q) Terminologies

**Quadrate:** Square.

**Quadri:** Prefix meaning four.

**Quinate:** With five nearly similar structures from a common point.

**Quinque:** Prefix meaning five.

# Glossary of Alphabet (R) Terminologies

**Raceme:** A single stemmed inflorescence with flowers on individual stalks along a stem. The bottom flowers open first as the raceme continues to elongate. Snapdragon and Delphinium flowers grow on racemes.

**Racemose:** Raceme like or bearing racemes.

**Rachilla:** A small rachis, a secondary axis of a multiply compound leaf.

**Rachis:** The main stalk of a flower cluster or of a compound leaf, also that part of a fern frond stem that bears the leaflets.

**Radial:** Symmetric when bisected from any angle.

**Radiate:** Describing a flower head in the *Asteraceae* that contains both ray and disk flowers.

**Radical:** Belonging to or proceeding from the root.

**Radicant:** Rooting from the stem.

**Radicle:** Initial root determined cells. Root apical meristem.

**Ramose:** Branching or branchy.

**Rank:** A vertical row usually of leaves or bracts that can be either opposite or alternate.

**Receptacle:** The end of the pedicel that joins to the flower was the different parts of the flower are joined together, also called the torus. In *Asteraceae* the top of the pedicel upon which the flowers are joined.

**Recumbent:** Leaning or reposing upon the ground.

**Recurved:** Curved backwards or outwards.

**Reflexed:** Abruptly bent or curved downward.

**Regular:** Describes a flower with petals or sepals all of equal size and shape, *i.e.* radially symmetrical or capable of being divided into mirror images on either side of any plane that passes through the center.

**Relative humidity:** The measurement of the amount of moisture in the atmosphere.

**Reniform:** Kidney shaped or rounded with a notch at the base.

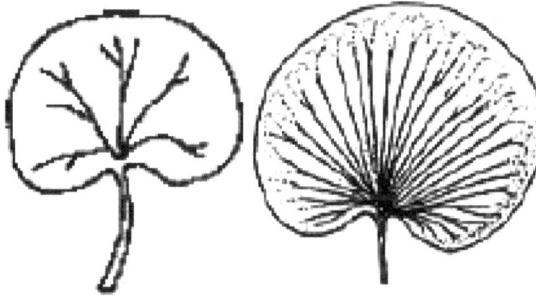

**Repand:** An undulating margin, less strongly wavy than sinuate.

**Repent:** Creeping.

**Reproductive:** Buds with embryonic flowers.

**Reticulate:** Having a netted pattern.

**Retrorse:** Bent backward or downward, reflexed.

**Retuse:** Having a rounded apex with a shallow notch.

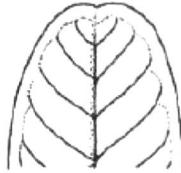

**Revolute:** Having the margins inrolled toward the underside.

**Rhizautoicous:** Male inflorescence attached to the female stem by rhizoids.

**Rhizome:** A horizontally orientated, prostrate stem with reduced scale-like leaves, normally growing under ground but also at the soil surface. Also produced by some species that grow in trees or water.

**Rhombic:** With the shape of a diamond.

**Root ball:** The network of roots along with the attached soil, of any given plant.

**Root bound:** A condition which exists when a potted plant has out grown its container. The roots become entangled and matted together, and the growth of the plant becomes stunted. When re potting, loosen the roots on the outer edges of the root ball, to induce them to once again grow outward.

**Root Hairs:** Very small roots, often one cell wide, that do most of the water and nutrient absorption.

**Root stimulator:** A root stimulator is a solution of specific nutrients formulated to encourage the growth and development of the roots of plants. Using root stimulator on newly planted and transplanted landscape additions will help them become established more quickly. A healthy and substantial root system will allow plants to flourish and resist adverse conditions such as extreme heat, cold and drought.

**Rooting hormone:** A powder or liquid growth hormone, used to stimulate a plant cutting to send out new roots from a stem node.

**Rootstock:** The underground part of a plant normally referring to a caudex or rhizome.

**Rosette:** A cluster of leaves with very short internodes that is crowded together, normally on the soil surface but sometimes higher on the stem.

**Rostrate:** Having a beak or beak like form.

**Rosulate:** Arranged into a rosette.

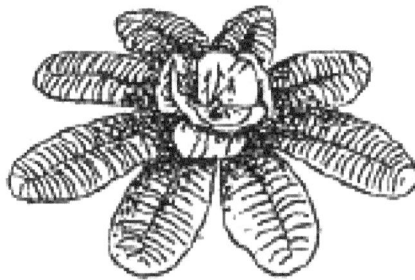

**Rotate:** A rotate corolla is wheel-shaped with a short tube and a wide horizontally flaring limb.

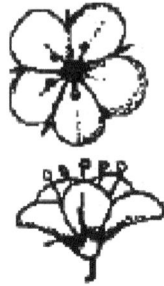

**Rotundifolius:** With round leaves.

**Rubescent:** Becoming red or reddish.

**Rubiginous:** Rust colored.

**Ruderal:** Growing in disturbed habitats, weedy.

**Rudiment:** An imperfectly developed organ, a vestige.

**Rufous:** Reddish brown.

**Rugose:** Wrinkled or bumpy.

**Runcinate:** Sharply incised or pinnatifid with the segments facing backwards.

**Runner:** A slender stem growing out from the base of some plants, which terminates with a new off set plant. The new plant may be severed from the parent after it has developed sufficient roots.

# Glossary of Alphabet (S) Terminologies

**Saccate:** Sac shaped or pouch-shaped.

**Sagittate:** Arrowhead shaped, with two retrose basal lobes.

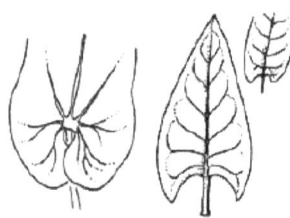

**Salient:** Projecting outward.

**Salverform:** With a slender tube abruptly expanded into a rotate limb.

**Samara:** Dry fruit with wings that do not open when mature, as in maple trees.

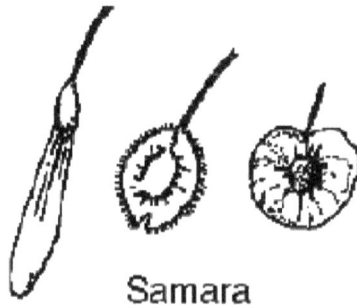

Samara

**Sanguineous:** Blood red.

**Sapid:** With an agreeable taste.

**Saponaceous:** Soapy.

**Saprophytic:** Deriving food from dead or decaying organic material in the soil and usually lacking in chlorophyll.

**Sarcocaulis:** With fleshy stems.

**Sarcotesta:** A fleshy seed coat.

**Sauveolent:** Fragrant.

**Saxatile:** Growing among rocks or in rocky, arid situations.

**Scaberulent:** Slightly scabrous.

**Scabrous:** Rough to the touch.

**Scalariform:** Ladder like.

**Scale:** A greatly reduced leaf or other outgrowth on a plant surface.

**Scandent:** A stem that climbs.

**Scape:** Leafless flowering stem arising directly from the ground.

**Scarification:** Scratching or nicking of a seed's shell to facilitate germination.

**Scarify:** To roughen, score or scrape the hard, outer coating of a seed to assist in the absorption of moisture before germination, a process that many desert wash seeds require.

**Scarious:** Thin, dry, membranous and more or less translucent.

**Scion:** A short length of stem, taken from one plant which is then grafted onto the root stock of another plant.

**Scissile:** Splitting easily.

**Sclerophyllous:** With stiff, firm leaves.

**Sclerotesta:** A hard seed coat.

**Scorpioid:** Describing a coiled inflorescence.

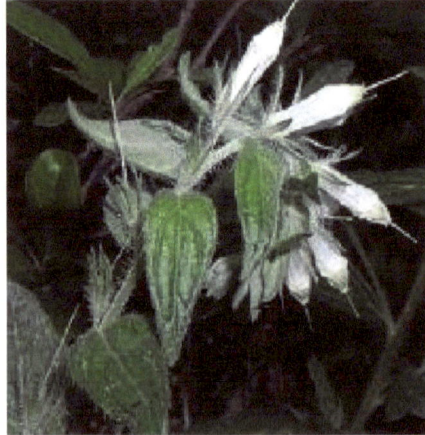

**Screening:** Screening plants produce dense growth and significant height, allowing them to provide privacy in urban settings or establish wind breaks in large open spaces. Taller varieties also furnish a vertical element to landscapes, drawing the eye upward. Use screening plants to divide large spaces and create cozy secluded hide a ways, block street noise, or hide un attractive views. Tall screens can also afford protection from the wind which can substantially reduce energy consumption in the winter time.

**Scurfy:** Covered with small scale-like or bran-like particles or projections.

**Sebaceous:** Tallow or fatty.

**Secondary:** Roots forming off of the primary root, often called branch roots.

**Secund:** Borne from only one side of an axis.

**Semi dwarf:** Semi-dwarf plants are ones which appear significantly smaller than members of the same species. This is often achieved by grafting a stem of a desirable variety to a root stock of a different variety. The difference in root stock and top allows the plant to grow only to a small percentage of its usual height and width, allowing it to be grown in a smaller space than its full sized counterpart.

**Semi erect:** Not growing perfectly straight.

**Semi:** Prefix meaning half.

**Semiamplexicaul:** The leaf base wraps around the stem, but not completely.

**Sepal:** A single segment of a divided calyx.

**Septate:** Divided by one or more partitions.

**Septicidal:** A capsule longitudinally dehiscent through the ovary wall at or near the center of each septa, preserving each locule as an intact entity.

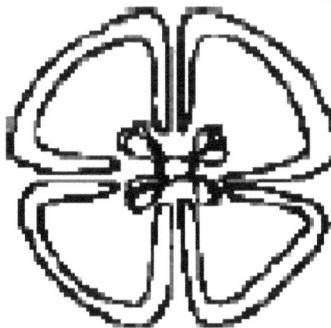

**Septum:** Any kind of a partition, specifically the wall between chambers in a compound ovary.

**Seriate:** Arranged in rows or series.

**Sericeous:** Covered with long, soft, straight, appressed hairs giving a silky appearance.

**Serpentine:** Refers to soils that are low in calcium and high in magnesium and iron, derived from greenish or gray-green rocks that are essentially magnesium silicate, other characteristics of which are a high nickel and chromium content, and a low content of nutrients such as nitrogen.

**Serrate:** Having sharp, forward-pointing teeth on the margin.

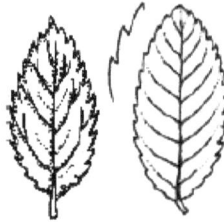

**Serrulate:** Serrate with very small teeth.

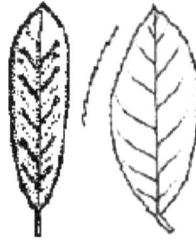

**Sessile:** Attached directly and without a petiole, pedicel or other type of stalk, said of either leaves or flowers.

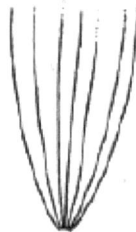

**Setaceous:** Bristle like.

**Setose:** Covered with bristles.

**Shearing:** Shearing is the cutting of the tips of the branches of a shrub, either manually with hand held pruners or with a motorized tool, to achieve a desired shape.

**Sheath:** The proximal portion of a grass leaf usually surrounding the stem, a leafy tubular structure usually on sedge or grass that envelopes the stem.

**Shrubs:** Shrubs are woody plants usually with multiple stems arising from or near their bases.Shrubs will not develop a bare trunk like a tree, however, some large shrubs can be pruned into tree-form by removing all but one straight main stem.

**Sigmoid:** Double-curved, S-shaped.

**Silicle:** A fruit similar to a silique, but much shorter, not much longer than wide.

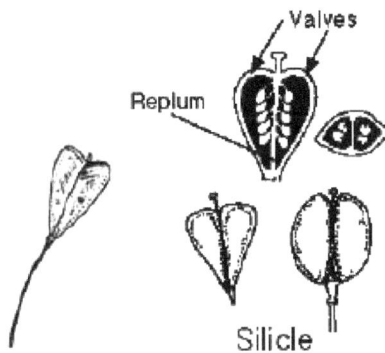

Silicle

**Silique:** A type of capsule found in the Brassicaceae, either half of which peels away from acentral, transparent, dividing membrane.

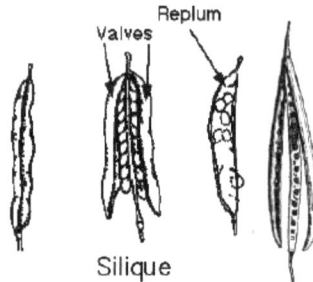

**Simple leaf:** A leaf that has one part, not subdivided into leaflets.

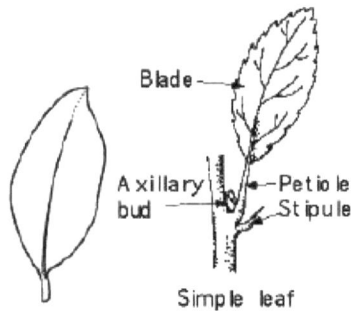

**Single:** One flower per stem or the flowers are greatly spread-apart as to appear they do not arise from the same branch.

**Sinistrorse:** Turned to the left or spirally arranged to the left.

**Sinuate:** Strongly or deeply wavy, usually referring to a leaf margin.

**Sinus:** The space or division, usually on a leaf, between two lobes or teeth.

**Soil conditioner:** Soil conditioners are products used to change the structure and fertility of existing or native soil. They are incorporated into the soil to help improve the vigor and overall performance of plants. Examples of soil conditioners include compost, manures and mulches along with many others.

**Solitary:** Same as single, with one flower per stem.

**Sori:** Clusters of spore sacs on a fern frond.

**Spadix:** A floral spike or head in which the flowers are borne on a fleshy axis.

**Spathe:** A large bract or pair of bracts subtending and usually partially enclosing an inflorescence.

**Spathulate:** Spoon shaped, gradually widening to a rounded apex.

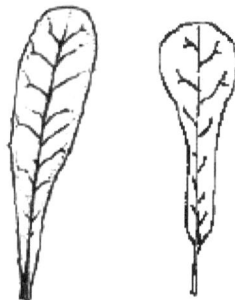

**Spear shaped:** Pointed with barbs.

**Specific epithet:** The second part of a scientific name which identifies the species.

**Specimen:** A specimen plant is one which captures attention with its unique structure, outstanding coloration, significant size or a combination of the three.

Specimens are usually planted singly within a landscape setting to draw attention to their uniqueness.

**Sphagnum:** A moss which is collected and composted. Most peat moss is composed primarily of sphagnum moss. This moss is also packaged and sold in a fresh state, and used for lining hanging baskets and air layering.

**Spicule:** A short, pointed, epidermal projection.

**Spike:** When flowers arising from the main stem are without individual flower stalks. The flowers attach directly to the stem.

**Spikelet:** In grasses, the smallest aggregation of florets plus any subtends glumes.

**Spine:** An adapted leaf that is usually hard and sharp and is used for protection, and occasionally shading of the plant.

**Spinescent:** Bearing a spine or spine like point.

**Spinose:** Having a stiff and tough acuminate tip.

**Spinulose:** Bearing very small spines.

**Sporangium:** A spore-case or sac in which spores are produced in a fern.

**Spore:** Spores are the reproductive cell structure of ferns, fungi and mosses. Fern spores develop inside small green capsules on the underside of the fronds, called sporangia. These types of plants do not form flowers nor produce seeds.

**Spp:** Abbreviation for species.

**Spumose:** Foamy or frothy.

**Spurred:** A hollow extension of a petal or sepal such as characterizes the larkspurs, and which often produces nectar.

**Squamate:** Having or producing scales.

**Squarrose:** Having spreading, recurved tips.

**Ssp:** Abbreviation for subspecies.

**Staked:** Plants are staked when a support is needed for upright growth. If a plant's natural habitis to trail or produce vines, it may be staked for guidance toward a structure or wall. Young trees are often staked in areas of high wind to prevent excessive swaying.

**Staking:** The practice of driving a stake into the ground next to, and as a support for a plant.When attaching the plant to the stake, is sure that it is tied loosely so it doesn't strangle the stem. When staking a potted plant, the stake should be set into the planter before the plant is added.

**Stamen:** The male or pollen-bearing organ of a flower composed of filament and anthers.

**Staminate:** Describing a male flower that contains one or more stamens but no functional pistils.

**Staminode:** A sterile stamen or other nonfunctional structure occupying the position and having the overall appearance of a stamen.

**Staminodial:** Concerning a sterile stamen, flowers with sterile stamens.

**Standard:** Standard size plants are either grown on their own roots or grafted onto a different root stock. Either method should not limit the plant from reaching the mature height for its species.

**Stellate:** Star like with radiating branches and often referring to the pattern of hairs on the surface of a leaf.

**Stem:** Vascular tissue that provides support to the plant, The main upward growing axis of a plant which bears the leaves and flowers.

**Stenopetalous:** With narrow petals.

**Stenophyllous:** With narrow leaves.

**Stigma:** The terminal portion of a pistil, which receives the pollen.

**Stipe:** The portion of a fern frond below the rachis, *i.e.* below where the leaflets are attached.

**Stipels:** Paired scales, spines, glands, or blade-like structures at the base of a petiolule.

**Stipitate:** Borne on a stipe or stalk.

**Stipule:** An appendage at the base of a petiole, usually in pairs.

**Stolon:** A branch that forms near the base of the plant and grows horizontally, and roots produce new plants at the nodes or apex, a horizontally growing stem similar to a rhizome, produced near the base of the plant. They spread out above or along the soil surface, roots and new plants develop at the nodes or ends, *e.g.*Strawberry plants.

**Stoloniferous:** Plants produce stolons.

**Stomata:** A small pore or opening on the surface of a leaf through which gaseous exchange takes place, *i.e.* the diffusion of carbon dioxide, oxygen and water vapor.

**Stone:** The hard, woody endocarp enclosing the seed of a drupe.

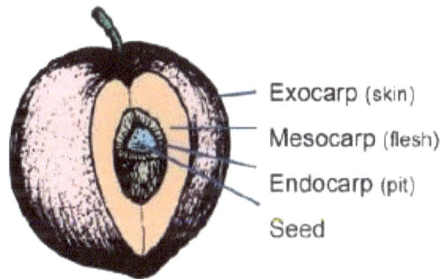

Exocarp (skin)

Mesocarp (flesh)

Endocarp (pit)

Seed

**Stramineus:** Straw colored.

**Stratification:** A process used to break the dormancy of a seed. This usually requires that the seeds be placed in a moistened rooting medium and kept in the refrigerator or freezer for a designated length of time.

**Striate:** With fine longitudinal lines or ridges.

**Striated:** Marked by a series of lines, grooves, or ridges.

**Strict:** Very straight and upright.

**Strigose:** Covered with rough, stiff, sharp hairs that are more or less parallel to a particular surface.

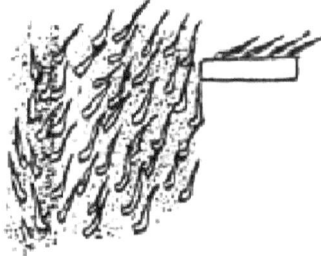

**Strobilus:** Inflorescences that is characterized by imbricated bracts or scales such as are borne on the ephedra.

**Style:** The narrowed portion of a pistil between and connecting the ovary and the stigma.

**Sub Tropics:** The subtropics are the areas between the tropical equatorial region and the cooleror more temperate regions of the planet. Many beautiful and durable landscape plants can be successfully grown in the subtropics. The subtropical locations near oceans or large bodies of water are the places of origin for some of the most colorful landscape plants.

**Sub:** Prefix meaning under, slightly, some what or almost.

**Suberose:** Having a corky texture.

**Subspecies:** A group of plants within a species that has consistent, repeating, genetic and structural distinctions.

**Subtend:** To occupy a position below and adjacent.

**Subulate:** Awl shaped with a tapering point.

**Succubous:** Describing the arrangement of leaves of a liverwort, contrast with incubus.

**Succulent:** Fleshy, juicy and thickened.

**Sucker:** A growth originating from the root stock of a grafted plant, rather than the desired part of the plant. Sucker growth should be removed, so it doesn't draw energy from the plant.

**Suffrutescent:** Some what shrubby, or shrubby at the base.

**Suffruticose:** Low shrubby, with the lower part of the stem woody and the upper part herbaceous.

**Suffused:** Tinted or tinged.

**Sulcate:** Grooved or furrowed.

**Summer annual:** Plant with seeds germinating in spring or early summer and completing.flowering and fruiting in late summer or early fall.

**Superior ovary:** One that is located above the perianth and free of it.

**Superphosphate:** Super phosphate is a fertilizer containing only the major nutrient phosphorus.It is an excellent fertilizer for a wide variety of plants including flowers, bulbs, perennials, roses and vegetables.

**Surcurrent:** Extending upward from the point of insertion, as a leaf base that extends up along the stem.

**Surficial:** Growing near the ground, or spread over the surface of the ground.

**Suture:** The seam along which the fruit opens, normally in most fruits it is where the carpel or carpels are fused together.

**Swale:** A depression or shallow hollow in the ground, typically moist.

**Sword shaped:** Long, thin, pointed.

**Sympatric:** Growing together with, or having the same range.

**Sympetalous:** Having the petals more or less united.

**Syn:** Prefix meaning united.

**Synandrous:** The anthers are connected, united.

**Syncarpous:** The gynoecium comprises one pistil.

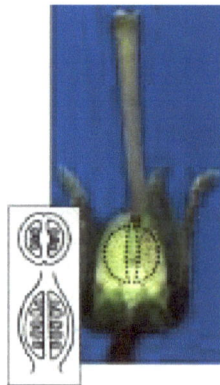

**Syngenesious:** The anthers are united into a tube, the filaments are free.

**Synoecious:** Having male and female flowers in the same flower head.

**Synoicous:** Male and female sex organs on the same gametophyte but are not clustered.

**Synsepalous:** Having the sepals more or less united.

**Systemic:** A chemical which is absorbed directly into a plants system to either kill feeding insects on the plant, or to kill the plant itself.

# Glossary of Alphabet (T) Terminologies

**Tannin bearing:** Tannin producing glands found in various parts of the plant, presumably protective in some structures.

**Tap root:** The main thick root growing straight down from a plant, the primary root continuing the axis of plant downward often quite deeply in to the ground.

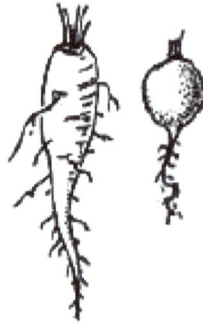

**Taxon:** Any group of plants occupying a particular hierarchical category, such as genus or species.

**Taxonomy:** The study of identification, nomenclature and classification of organisms is called as Taxonomy, Taxonomy is actually the study of relationships of organisms with their ancestors by evolutionary sequences.

**Tender plants:** Plants which are unable to endure frost or freezing temperatures.

**Tendril:** A thigmotropic organ which attached a climbing plant to a support, a portion of a stemor leaf modified to serve as a holdfast to other objects.

**Tenuous:** Slender or thin.

**Tepal:** A collective term for sepals and petals, used when they cannot be easily differentiated.

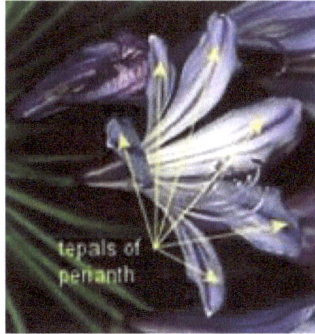

**Terete:** Circular in cross section.

**Terminal bud:** Bud at the tip or end of the stem.

**Tesselate:** Marked by a pattern of polygons, usually rectangles.

**Tessellated:** Color arranged in small squares, so as to have some resemblance to a checkered pavement.

**Testa:** The seed coat; develops from the integuments after fertilization.

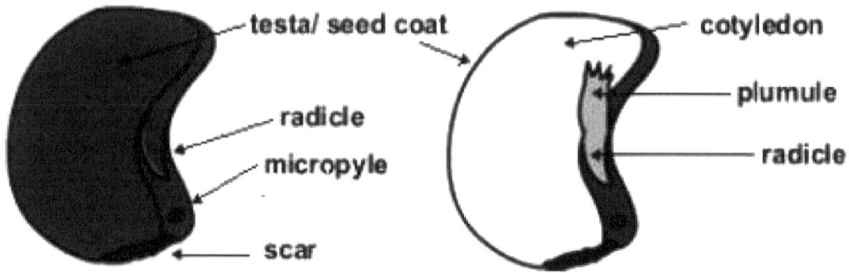

**Tetra:** Prefix meaning four.

**Tetracyclic:** Four whorled.

**Tetrad:** Pollen grains in clusters of four.

**Tetradynamous:** With stamens in two groups, usually four long and two short.

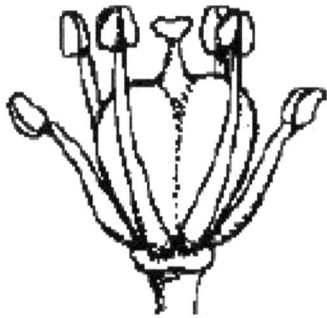

**Tetragonal:** Four angled.

**Tetrahedral:** Having the form of a tetrahedron.

**Tetralocular:** Four locular.

**Tetramerous:** Whorl with four members.

**Tetrandrous:** With four stamens.

**Tetrastichous:** Leaves or other structures in four rows.

**Thatch:** The layer of dead stems that builds up under many lawn grasses. Thatch should be removed periodically to promote better water and nutrient penetration in to the soil.

**Thinning:** Removing excess seedlings, to allow sufficient room for the remaining plants to grow. Thinning also refers to removing entire branches from a tree or shrub, to give the plant a more open structure.

**Thorn:** A short, stiff, sharp-pointed end.

**Throat:** In some corollas with fused petals, the point of juncture between the tube and limb, a some what difficult point to distinguish.

**Tiller:** In grasses the young vegetative shoots.

**Tomentose:** Wooly, with long, soft, matted hairs.

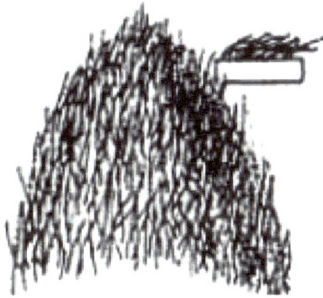

**Toothed:** Having small lobes or points along the margin as on a leaf.

**Top dress:** To evenly spread fertilizers or other soil amendments over the surface of the soil.

**Topiary:** A method of pruning and training certain plants in to good quality soil at nurseries and garden centers.

**Torose:** Cylindrical with contractions at intervals.

**Tortuous:** Twisted or bent.

**Trailing:** Trailing plants produce predominantly horizontal growth with little or no strongly upright branches. These plants lend themselves ideally for foreground planting, massing and as ground cover.

**Translator:** A structure uniting the pollinia in *Asclepiadaceae* and *Orchidaceae*.

**Transpiration:** The release of moisture through the leaves of a plant, emission of water vapor from the leaves.

**Transplanting:** The process of digging up a plant and moving it to another location.

**Transverse:** At a right angle to the longitudinal axis of a structure.

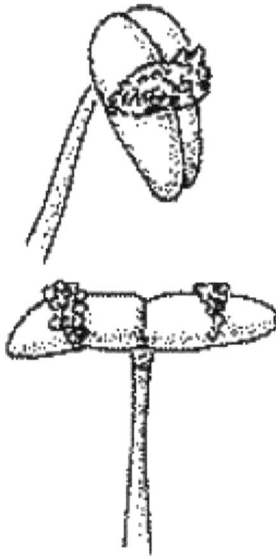

**Trees:** Trees are plants having wood with tall stem. *e.g.* Mangifera *indica, Ficus religiosa.*

**Tri:** Prefix meaning three.

**Triad:** A cluster of three, as spikelets of Hordeum or Hilaria.

**Triandrous:** Having three stamens.

**Tricarpellate:** Three carpellate.

**Trichome:** A hair like outgrowth from the epidermis.

**Trichotomous:** Three forked.

**Tricyclic:** Three whorled.

**Tridynamous:** With stamens in two equal groups of three.

**Trifid:** Three cleft to about the middle.

**Triflorous:** Three flowered.

**Trifoliate:** Divided in to three leaflets.

**Trifoliolate:** With three leaflets.

**Trifurcate:** Divided in to three forks or branches.

**Trigonous:** Three angled.

**Triheteranthous:** Having different states in three different sets of flowers, only one state present in each set.

**Triheterophytous:** Having different states in three different sets of plants, only one state present in each set.

**Trilete:** Basically tetrahedral, but often appearing round or triangular, with three scar lines forming a Y.

**Trimerous:** Whorl with three members.

**Trimorphic:** Elaving three different shapes and sizes within the same species.

**Trinucleate:** Pollen containing three nuclie when shed.

**Trioecious:** Plants staminate, pistillate or perfect.

**Tripinnate:** Pinnately compound in which each leaflet is it self bipinnate.

Tripinnate

**Triquetrous:** Three-angled with the sides usually concave.

**Tristichous:** Leaves or other structures in three rows.

**Tropical plants:** A tropical plant is one which comes from a region which does not experience any freezing temperatures, thus allowing it to grow all year in its native environment.

**Tropics:** The tropics are areas in the form of a band with the equator at its center which encircles the earth. Plants from the tropics usually come from regions of consistently warm temperatures with moderate to heavy rainfall.

**Tropism:** The turning of a plant part such as a leaf in response to some external stimuli.

**True indusium:** An epidermal outgrowth protecting the sorus.

**Truncate:** With a base or apex appearing as if cut straight across.

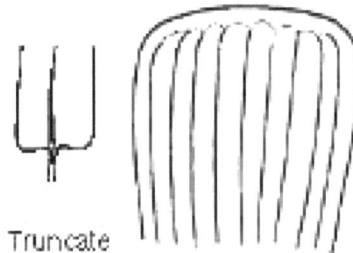

Truncate

**Tuber:** An underground stem which stores food and plant energy and from which a plant grows.

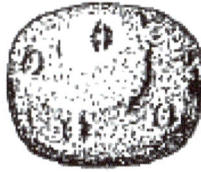

Tuber

**Tubercle:** A knob like projection.

**Tuberculate:** With a warty surface.

**Tubercules:** Silica deposits on the stem ridges, as in Equisetum.

**Tuberous:** Describes roots that are thick and soft, with storage tissue, typically thick round in shape.

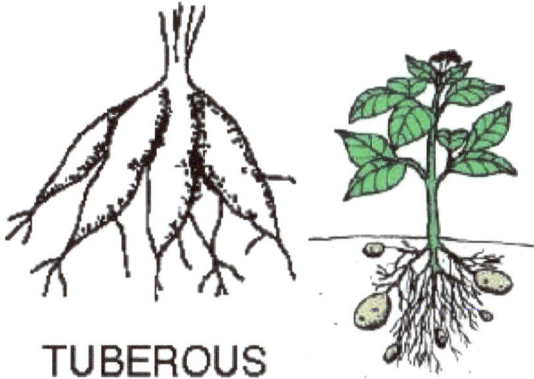

TUBEROUS

**Tubular:** The lower or narrower portion of a corolla or calyx.

**Tufted:** In a dense cluster.

**Tumescent:** Some what tumid, swelling.

**Tumid:** Swollen.

**Tunicate:** Having several concentric layers, such as in onions.

**Turbinate:** Shaped like a top or inverted cone.

**Turgid:** Swollen.

**Twining:** Climbing by coiling around some support.

150

# Glossary of Alphabet (U) Terminologies

**Umbel:** A mostly flat topped flower cluster in which individual flower stems radiate from a common point, like the ribs of an umbrella.

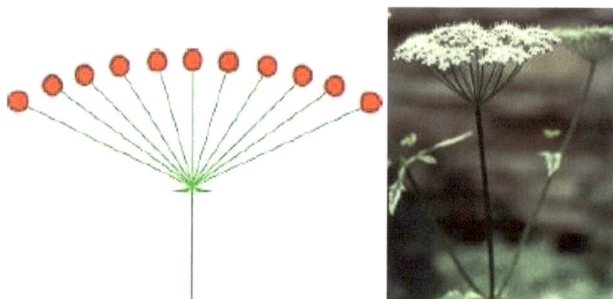

**Umbellet:** A secondary umbel in a compound umbel.

**Umbellulate:** In the form of or having the appearance of an umbel.

**Umbo:** Projection, with or without spine or prickle, on the apophysis of the cone scale.

**Umbonate:** Round with a projection in the center.

**Umbraculate:** Umbrella shaped.

**Unarmed:** Lacking thorns or prickles.

**Uncinate:** Hooked near the apex.

**Unctuous:** Greasy, oily.

**Undulate:** Margins shallowly and smoothly indented, wavy in a vertical plane.

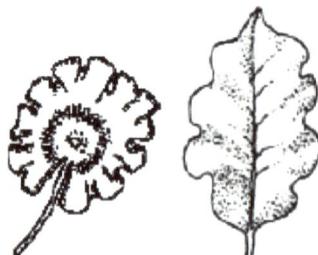

**Unguiculate:** Contracted at the base into a claw as a petal.

**Uni:** Prefix meaning one.

**Unifoliate:** With a single leaf.

**Unilocular:** Having a single locule in the ovary.

**Uninodal shoot:** Spring shoot developing from the terminal winter bud and producing only one internode with one whorl of branches at the end, the cones are sub-terminal at the end of the shoot. *e.g. Pinus resinosa.*

**Uniseriate:** Arranged in one row or series.

**Unisexual:** With only one sex in each flower, bearing either stamens or pistils.

**United:** Describes petals that are fused togather.

**Upright:** Upright plants produce a vertical branch which exceeds the length of their horizontal branching.

**Urceolate:** Urn shaped or pitcher like, Contracted at the mouth.

**Urent:** With erect usually long trichomes that produce irritation when touched.

**Utricle:** A small thin walled single seeded more or less bladdery inflated fruit.

**Uva:** Grape like berry formed from a superior ovary.

# Glossary of Alphabet (V) Terminologies

**Vaginate:** Provided with or surrounded by a sheath.

**Vallecular canal:** A canal beneath a stem groove.

**Valvate:** Having margins of adjacent structures touching edge only.

**Valve:** One of the parts or segments into which a dehiscent fruit splits.

**Valve:** A cardinal organ part that is a part of a plant structure that splits apart when the structure dehisces.

**Valvular:** Anther openings by valves or small flaps.

**Varicose:** Swollen or enlarged in places.

**Variegated:** The color disposed in various irregular sinuous, spaces.

**Vascular bundle:** A portion of vascular tissue that is a unit strand of the vascular system and has as part xylem or phloem.

**Vascular bundles:** A strand of woody fibers and associated tissues.

**Vascular cambium:** A cambium that is located between and gives rise to secondary xylem and secondary phloem.

**Vascular leaf anlagen:** A phyllome anlagen that will give rise to a vascular leaf primordium and is part of a peripheral zone of a vegetative shoot apical meristem.

**Vascular leaf meristematic apical cell:** A leaf meristematic apical cell that is part of the leaf apex of a vascular leaf.

**Vascular leaf primordium:** A phyllome primordium that develops from leaf anlagen and is part of a vegetative shoot apex and is committed to the development of a vascular leaf.

**Vascular leaf:** A leaf in a vascular plant.

**Vascular shoot axis meristematic apical cell:** A shoot axis meristematic apical cell at the tip of a shoot apex in a shoot system that has as part vascular tissue.

**Vascular system:** A maximal portion of vascular tissue in a whole plant collective plant structure multi-tissue plant structure or cardinal part of multi-tissue plant structure.

**Vascular:** Containing xylem, the water and mineral conducting tissue, and phloem, the food conducting tissue.

**Vegetative bud:** A bud that develops into a shoot system that has as organ parts only vegetative organs.

**Vegetative cell:** A plant cell that is the larger cell of a male gametophyte in seed plants. It does not divide further and develops into a pollen tube cell.

**Vegetative frond:** Frond lacking sporangia.

**Vegetative shoot apex:** A shoot apex that has as part a vegetative shoot apical meristem.

**Vegetative shoot apical meristem:** A shoot apical meristem that gives rise to the apical growth of vegetative tissues and organs.

**Vegetative:** Buds containing embryonic leaves.

**Vein:** The externally visible vascular bundles, found on leaves, petals and other parts.

**Veinlet:** A small vein.

**Velamen:** A multiseriate epidermis found in aerial roots of some monocots. Most of its cells are dead and store water like a sponge.

**Velum:** The membranous flap covering the sporangium

**Velutinous:** Covered with dense straight long and soft trichomes.

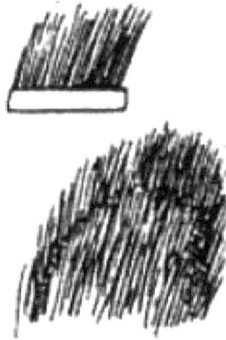

**Venation:** The arrangement of veins on leaf.

**Venter:** A cardinal organ part that is the enlarged basal part of an archegonium and has an archegonium egg cell located in it.

**Ventral canal cell:** A plant cell that develops from the archegonium central cell and lies at the base of the archegonium neck above the archegonium egg cell in the venter.

**Ventral:** Pertaining to the surface nearest to the axis, the upper surface of the leaf.

**Ventricose:** Inflated or swollen unequally on one side.

**Ventristipular:** On ventral side of stipule.

**Vermicular:** Worm shaped or worm like appearance.

**Vermiculite:** A sterile soil amendment created when the mineral has been heated to the point of expansion, like popcorn, a good addition to container potting mixes vermiculite retains moisture and air within the soil.

**Vermiform:** Worm shaped.

**Vernal:** Appearing in the spring.

**Vernation:** The arrangement of leaves in an un opened bud.

**Verrucose:** With a wart surface, with low rounded bumps.

**Versatile:** Reffering to anther which attaches at or near its middle and is able to turn freely on its support.

**Versicolor:** Having various colors.

**Verticil:** Flowers arranged in whorls at the nodes.

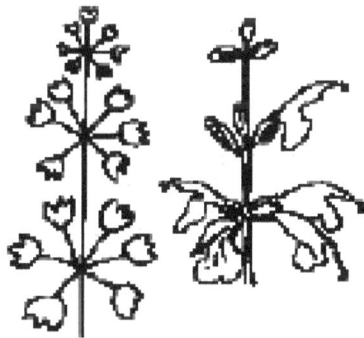

**Verticillaster:** A whorled collection of flowers around the stem, the produced in rings at intervals up the stem. As the stem tip continues to grow more whorls of flower are produced.

**Verticillate:** Arranged in whorls.

**Vesicle:** A bladder or cavity.

**Vespertine:** With flowers opening in the evening or night; appearing or expanding in the evening.

**Vexillate:** Having one structure larger than others which is folded over smaller enclosed structures.

**Villose:** Covered with fine long hairs that are not mated.

**Villosity:** Villous indument.

**Villosulous:** Minutely Villous.

**Villous:** Covered with long, soft and crooked trichomes.

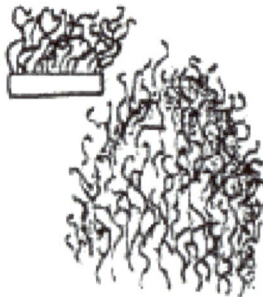

**Vinaceous:** Wine colored.

**Vine:** A is considered a plant that will grow to an indefinite height and or width while at same time depending on another plant or surface for support.

**Violaceous:** Violet colored.

**Virgate:** Wand-like, slender erect grpwing stem with many leaves or very short branches.

**Viscid:** Sticky or glutinous.

**Vitreous:** Transparent.

# Glossary of Alphabet (W) Terminologies

**Wanting:** Lacking, absent.

**Water excreting:** Specialized vein openings through which water is lost under certain atmospheric conditions. *e.g* hydathods.

**Weed:** A plant that intrudes where it is not wanted a plant that vigorously colonizes disturbed areas.

**Whole plant:** A plant structure which is a whole organism.

**Whorl:** A circle of three or more structures radiating outward from the same node.

**Whorled:** A collection of three or more leaves or flowers that arise from the same point.

Whorled

**Wing petal:** One of two lateral petals that is adjacent to the banner petal of a *papiliona ceouscorolla*.

**Wing:** The term used for the lateral petals of the flowers. *e.g.Fabaceae*.

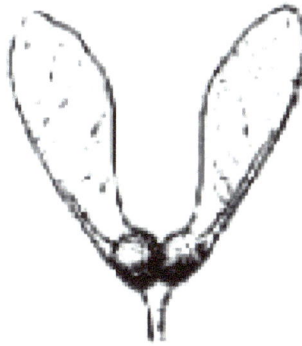

**Winged nut:** Nut enclosed in a wing like bract, as in *carpinus*.

**Winged schizocarp:** A schizocarp is a dry or fleshy fruit derived from two to many carpellate gynoeciums that breaks into one or few seeded segments at maturity, a winged schizocarp is described as a samara like schizocarp.

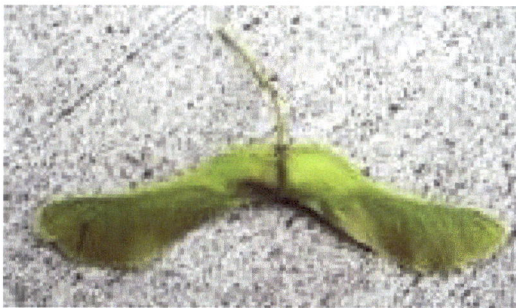

**Winter annual:** Plants with seeds germinating in late summer or fall and completing flowering and fruiting in spring or summer.

**Woody perennial:** The shrubs, trees and some vines with shoot systems that remain alive above the soil surface from one year to the next.

**Woody:** Secondary growth laterally around the plant so as to form wood.

**Woolly:** Having soft wool like hairs.

# Glossary of Alphabet (X) Terminologies

**Xanthic:** Yellowish in color.

**Xenia:** The effect of pollen on seeds and fruits.

**Xeric:** Pertaining to arid and desert conditions, implying a minimal water supply throughout most of the year.

**Xero:** Prefix meaning dry.

**Xerophytic plants:** Plants adapted to dry or arid conditions, places where fresh water is scarce where water absorption is difficult due to an excess of dissolved salts.

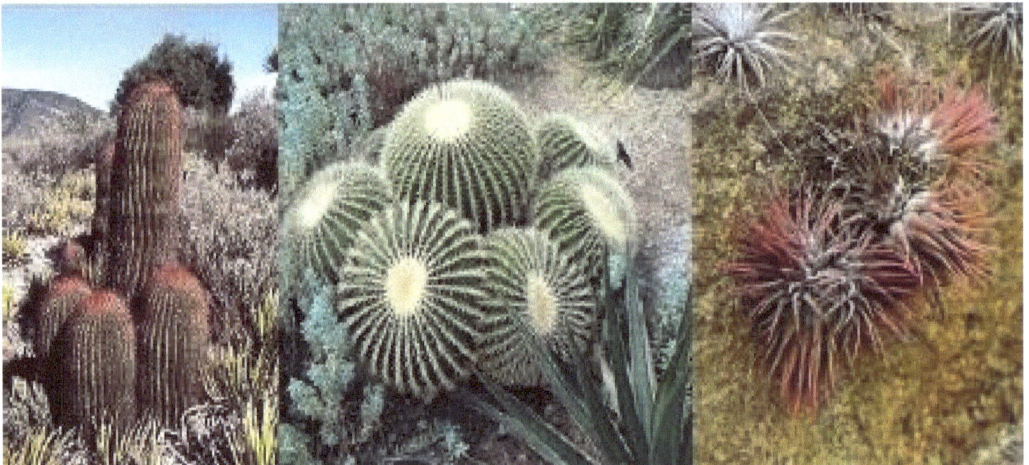

**Xylem fiber cell:** A fiber cell that is part of a portion of xylem tissue.

**Xylem pole pericycle cell:** A pericycle cell that is adjacent to the protoxylem of a vascular bundle.

**Xylem sap:** A plant sap that is an aqueous solution which may contain mineral elements, nutrients, and plant hormones, and is transported from the root system toward the leaves through the tracheary elements of the xylem.

**Xylem vessel member:** A tracheary element that is part of a xylem vessel and has as parts perforation plates.

**Xylem vessel:** A portion of xylem tissue that has as parts a tube-like series of vessel members the common walls of which have perforations.

**Xylem:** The water conducting tissue of vascular plants.

**Xylocarp:** A hard woody fruit such as the coconut.

# Glossary of Alphabet (Y) Terminologies

**Yeast:** Unicellular fungus whose colonies resemble those of bacteria, known as the microorganisms that make bread rise.

**Yield plateau:** A temporary stable state in yield reached in the course of increased production.

**Y-linked gene:** A gene found on the Y chromosomes.

# Glossary of Alphabet (Z) Terminologies

**Zein:** A seed protein of maize classified as prolamin, it is low in tryptophan and lysine.

**Zonate:** Marked or colored in circular rings or zones.

**Zoned:** The same as ocellated but the concentric bands more numerous.

**Zonocaulous:** With branches intermittently spaced along main stem.

**Zoophlious:** Animal pollinated.

**Zoospore:** A flagellated spore.

**Zygomorphic:** One axis of symmetry running down the middle of the flower so the right and left halves reflect each other.

**Zygomorphy:** The type of symmetry that most irregular flowers have with the upper half of the flower unlike the lower half, the left and right halves.

**Zygosporangium:** In fungi fused gametangia in conjugation.

**Zygospore:** The protective structure that results when the wall surrounding a zygote thickness.

**Zygote:** The cell resulting from fusion of gametes.

**Zygote:** The product of the fusion between one of the pollen sperm nuclei and the egg cell of the female gamete at fertilization. Following a number of mitotic divisions, the zygote differentiates into the embryo.

**Zygotene:** A stage at prophase I of meiosis during which homologous chromosomes pair with each other.

**Zygotic plant embryo:** A plant embryo that forms as a result of the fusion of gametes.

# BIBLOGRAPHY

Available from: http://www.plantontology.org/ontology/index.html.

Available from: http://www.ibiblio.org/botnet/glossary/syschar.html.

Available from: http://www.bio.miami.edu/dana/226/226F09_21.html.

Available from: http://botany.csdl.tamu.edu/FLORA/201Manhart/Homepage.html.

Available from: https://www.integratedbreeding.net/./plant.and./index-id=003.php.html.

Core, Earle L. and Nelle P. A. (1958). *Woody Plants in Winter: A Manual of Common Trees and Shrubs in Winter in the Northeastern United States and Southeastern Canada*. The Boxwood Press: Pittsburgh, Pennsylvania.

Dirr, M. A. (1998). *Manual of Woody Landscape Plants: Their Identification, Ornamental Characteristics, Culture, Propagation and Uses*, 5th ed. Stipes Publishing: Champaign, Illinois.

Harris, J. G. and Melinda W. H. (2001). *Plant Identification Terminology: An Illustrated Glossary*, 2nd ed. Spring Lake Publishing: Spring Lake, Utah.

Hickey, Mchael and Clive King. (2000). *The Cambridge Illustrated Glossary of Botanical Terms*. Cambridge University Press: Cambridge.

Hightshoe, G. L. (1978). *Native Trees for Urban and Rural America: A Planting Design Manual for Environmental Designers*. Iowa State University Research Foundation: Ames, Iowa.

Kiger, R. W. and Duncan M. P. (2001). *Categorical Glossary for the Flora of North America Project*. Hunt Institute for Botanical Documentation, Carnegie Mellon University: Pittsburgh, Pennsylvania.

Lawrence, George H. M. (1951). *Taxonomy of Vascular Plants*. The MacMillan Company: New York.

Radford, A. E., William C. D., Jimmy R. M. and C. Ritchie Bell. (1974). *Vascular Plant Systematics*. Harper & Row Publishers: New York, Evanston, San Francisco, London.

Raven, Peter H., Ray F. Evert, and Susan E. Eichhorn. (1999). *Biology of Plants*, 6th ed. Worth Publishers: New York.

Walters, D. R. and David J. Keil. (1996). *Vascular Plant Taxonomy*, 4th ed. Kendall/Hunt Publishing: Dubuque, Iowa.

Zomlefer, Wendy B. (1994). *Guide to Flowering Plant Families*. University of North Carolina Press: Chapel Hill, NC & London.